LEARN
Like a Pro

Also by Dr. Barbara Oakley

Learning How to Learn

Mindshift

A Mind for Numbers

Cold-Blooded Kindness

Pathological Altruism

Evil Genes

Hair of the Dog

Practicing Sustainability

Career Development in Bioengineering and Biotechnology

Also by Olav Schewe

Super Student

The Exchange Student Guidebook

LEARN
Like a Pro

Science-Based
Tools to Become Better
at Anything

DR. BARBARA OAKLEY
and **OLAV SCHEWE**

ST. MARTIN'S
ESSENTIALS
NEW YORK

First published in the United States by St. Martin's Essentials, an imprint of
St. Martin's Publishing Group

www.stmartins.com

Library of Congress Cataloging-in-Publication Data

Names: Oakley, Barbara A., 1955– author. | Schewe, Olav, author.
Title: Learn like a pro : science-based tools to become better at anything /
   Barbara Oakley and Olav Schewe.
Description: First edition. | New York : St. Martin's Essentials 2021. |
   Includes bibliographical references and index. |
Identifiers: LCCN 2020056422 | ISBN 9781250799371 (trade paperback) |
   ISBN 9781250799388 (ebook)
Subjects: LCSH: Learning, Psychology of. | Cognitive psychology. |
   Cognitive neuroscience. | Study skills.
Classification: LCC LB1060 .O25 2021 | DDC 370.15/23—dc23
LC record available at https://lccn.loc.gov/2020056422

Our books may be purchased in bulk for promotional, educational, or business
use. Please contact your local bookseller or the Macmillan Corporate and
Premium Sales Department at 1-800-221-7945, extension 5442, or by email at
MacmillanSpecialMarkets@macmillan.com.

First Edition: 2021

10 9 8 7 6 5 4 3

Contents

To Our Readers

Do you spend too much time learning with disappointing results? Do you find it difficult to remember what you read? Do you put off studying because it's boring and you're easily distracted?

This book is for you.

We're Olav Schewe and Barb Oakley, and both of us have struggled in the past with our learning. But we have found techniques to help us master material—*any* material. Building on insights from neuroscience and cognitive psychology, we will give you a crash course to improve your ability to learn, whether you're studying math, language, coding, karate, cooking, or anything else. You'll see *why* the strategies work because you'll see what's happening in the brain when you use them. No, this isn't a little book of miracles. But you will find that reducing frustration and improving your study success may sometimes feel miraculous.

In Olav's case, he always wanted to get good grades, but he couldn't, no matter how hard he studied. He almost gave up on his dreams, thinking he wasn't smart enough. But then he discovered that the key to better grades wasn't his own innate abilities, or the number of hours he put in—instead, it was *how* he studied. After he took a step back in his teens and made some adjustments in his learning techniques, he began to excel.

In the end, Olav, the former "slow" student with average grades, became the top student in his high school class. He went on to get a master's with distinction in business administration from the University of Oxford. And his book on how to study effectively, *Super Student,* became an international best seller—translated into more than a dozen different languages.

As for Barb, she flunked math and science all the way through high school. She was convinced she didn't have "the math gene." In her later twenties, however, she decided to start all over again with math, beginning with remedial high school algebra. Slowly, she improved in math and science. By applying powerful study methods learned during her language study at the United States Defense Language Institute, she succeeded. She's now a professor of engineering—she also teaches millions of students around the world in online courses such as "Learn like a Pro" and "Learning How to Learn," which is one of the world's largest massive open online courses. This goes to show that even when you believe you're genetically incapable of succeeding at a subject, that doesn't need to be true at all.

You might think, "I am terrible with numbers," or "I can't learn languages," or "I can't speak in front of crowds," when the reality is that you simply haven't learned how to do those things—*yet*. Instead, if you change your focus to finding techniques and methods for learning, you'll find that this little book puts marvelous insights at your fingertips. You'll find new solutions, often based on recent neuroscientific findings, that will allow you to move mental mountains and do what you never thought you could do.

Through their decades of writing, teaching, and research on learning, Olav and Barb have developed deep connections with experts from a vast array of disciplines. What you'll find in this little book are the very best of practical learning tools and insights synthesized from research in neuroscience, cognitive psychology, education, and many other fields. And it's all honed with feedback from hundreds of pro learners, many of whom themselves have gone through the trenches of learning difficult concepts and skills. Pro learners gradually add tools and techniques to their mental toolbox, and discover how to think more critically about their learning. That allows these learners to make the best use of their brains, whether those brains seem "naturally" geared toward learning or not.

This book will help you join the pro learning club. Welcome aboard!

LEARN
like a Pro

1

How to Focus Intently and Beat Procrastination

You're reading this book because, no matter what you're trying to learn, you want to make sure every minute of your studies counts. So let us start by giving you one of the simplest, most powerful mental tools in the world of study: the Pomodoro Technique.* This clever method will definitely help boost your concentration—we know this from research. Even if you already know the Pomodoro, you'll discover modern twists that can make the Pomodoro even more powerful. And there's much more in the chapters ahead that will be new to you!

The Pomodoro Technique

Use this approach to structure your study sessions:

1. **Sit down where you'll be studying or working and remove any possible distractors.** This means ensuring there are no pop-ups or extraneous open tabs on your

---

*This technique was invented by Italian Francesco Cirillo in the 1980s and named after his round tomato-shaped kitchen timer. "Pomodoro" means tomato in Italian.

computer, dings from your cell phone, or anything else that
could draw you off-task.

2. **Set a timer for 25 minutes.** You
   can use a mechanical or silent digital
   timer. You can also use the timer
   or an app on your phone. If you use
   your phone, place it out of sight and
   beyond arm's reach so you won't be
   distracted while focusing.

   **POPULAR
   POMODORO APPS**
   - focus booster (PC)
   - PomoDone
   - Forest
   - Toggl

3. **Dive in and study or work as intently as you can** for
   those 25 minutes. If your mind wanders (as it inevitably will),
   just bring your attention back to the task at hand. Most

4 Take a 5-minute break!

1 Remove all distractions

3 Focus intently on your work

2 Set a 25-minute timer

The Pomodoro Technique in four easy steps.

things can wait or be postponed for 25 minutes. If distracting thoughts come up that you feel like you should act on, write them down in a to-do list so you can tackle them after the Pomodoro session is over.

4. **Reward yourself** for about 5 minutes at the end of the Pomodoro session. Listen to your favorite song, close your eyes and relax, go for a walk, make a cup of tea, cuddle with your dog or cat—anything to let your mind comfortably flow free. It's also best to avoid checking your cell phone or email during this break time—more on why later.

5. **Repeat** as appropriate. If you want to study for 2 hours, you can do four Pomodoros with the break lasting roughly 5 minutes each time. If you have trouble getting yourself back to work when the break is done, set a timer for the break as well.

Sounds easy, doesn't it? It *is* easy. Sometimes your mind may struggle while doing a Pomodoro, but the reality is almost anyone can keep their focus for 25 minutes.

Why the Pomodoro Technique Works

You might wonder how something so simple can be so powerful. The reason is that the Pomodoro Technique captures important aspects of how your brain learns.

- Pomodoro-fueled bursts of focused attention give your brain practice in focusing without disruption, which is much needed in today's distraction-ridden mobile phone world.[1]

- Short mental breaks where you get away from focusing are ideal to allow you to transfer what you've just learned into long-term memory, clearing your mind for new learning.[2] You can't feel this process taking place, which is why you might tend to skip it—but *don't skip it*!

- Anticipation of a reward keeps you motivated throughout the Pomodoro.

- It's much easier to commit and recommit yourself to short bursts of dedicated study than to seemingly endless sessions.

- Your studies begin a pattern of focusing on the *process*— putting in certain amounts of dedicated time—rather than the *goal* or *outcome*. In the long run, having a good process in place is much more important than any one individual session or goal.

- When you even just think about something you don't like or don't want to do, it activates the insular cortex, causing a "pain in the brain." This pain diminishes after about 20 minutes of focus on the activity.[3] Twenty-five minutes is therefore perfect to get you into study mode.

The Pomodoro Technique is highly adaptable. If you get into the flow and find yourself wanting to continue past 25 minutes, that's okay. The length of the reward period is also flexible and can be longer than 5 minutes if your Pomodoro has gone longer than the usual 25 minutes. Just don't forget that taking a mental break is important. One analysis of data from a time-recording app found that highly productive workers work for an average of 52 minutes with a 17-minute break.[4] The key was that when these superstar workers focused, they *focused,* and when they took a break, they really took a break.

If you have nothing else to do once you've finished your Pomodoro, good. But if you have still more work to do, take a 5-minute break (set a "break" timer if you need to), then start the next Pomodoro. If you're doing a series of Pomodoros, try taking a longer, 10- to 15-minute break after every third or fourth Pomodoro you complete.

If you use the Pomodoro method to study new material, it's also wise to **spend at least some minutes of the Pomodoro looking away from what you're studying and trying to *recall* what you**

have just learned. As you will see in chapter 3, *recall* (also called *"retrieval practice"*) is one of the most powerful ways to both remember and understand new information.

Avoid Your Mobile Phone During Learning Breaks

Research by professors Sanghoon Kang and Terri Kurtzberg from Rutgers Business School has revealed that **using a mobile phone for a break does not allow your brain to recharge as effectively as the other types of breaks**.[5] They note: "As people are increasingly addicted to their cell phones, it is important to know the unintended costs associated with reaching for this device every spare minute. Although people may assume that it is not different from any other kind of interaction or break, this study shows that the phone might be more cognitively taxing than expected."

Mobile phones are also particularly distracting if you happen to be in face-to-face training or classes. One study found that "Students who were not using their mobile phones wrote down 62% more information in their notes, were able to recall more detailed information from the lecture, and scored a full letter grade and a half higher on a multiple choice test than students who actively used their mobile phones."[6] Even just having your phone near you while you're studying can be distracting—your brain is still tracking it if it knows it's close at hand.[7]

If you feel anxious without a phone, researchers have found you'll still be better off with the phone out of reach.[8] Leave your phone in a backpack, briefcase, or purse, or even back in the car. You'll be shocked at how much your focus improves.

Be Wary of Multitasking When Studying

Whenever you switch your focus to a new task, you activate information stored in your brain related to the new task.[9] When you then switch to a different task, for example, when checking your email or a text message, you activate a different set of information. This leaves what's called an *attention residue*—some leftover attention from your previous task that means your attention isn't fully on the new task. Frequent task-switching increases susceptibility to distraction, causes more errors, slows work, makes writing worse, diminishes learning, and causes forgetting. In short, it's bad news. One study by researchers at the University of Michigan found that cognitive performance fell by 30 to 40 percent when participants switched between tasks instead of completing one task before moving to the next.★[10] This is part of the magic of the Pomodoro Technique—it allows you to focus on *one* task without the interruptions that can cause you to multitask.

However, although researchers have focused on the dark side of multitasking, there is a brighter side—creativity. When you are focused on a task, you can become cognitively fixed on it. This reduces your ability to step back and take another approach or perspective. It seems that task-switching reduces cognitive fixation.[11] The question then arises, how often should you task-switch? There are no easy answers, because it depends so much on the task and how often you might get cognitively "stuck."

Earmuffs can be one of the best tools around to help you maintain focus. We recommend the 31 dB Peltor earmuffs (although large, they are virtually screaming-baby-proof) over more slender noise-canceling earmuffs.

If, on any particular day, you find yourself going off-task so often that your work is suffering, we recommend that you use the Pomodoro Technique. This will keep you on task. But if you're making good progress in your study session despite a few occasional peeks at something different, especially when you feel a bit stuck in your problem-solving or writing, you're probably doing just fine.

★There are a few people, about 2.5 percent of the population, who can efficiently switch their attention between different, complex activities. Odds are you are not one of these people—most brains just aren't set up that way. Medeiros-Ward, et al., 2015.

Set Up a Distraction-Free Environment

Taking occasional peeks at a distraction is one thing, but to avoid being frequently drawn completely off-task, you want to **find a place to study where distractions are eliminated or minimized**. Learning specialists recommend avoiding study in rooms where your friends or colleagues socialize—for example, a college dorm room or student or employee cafeteria. There can be too many interruptions. A quiet library or an isolated location can be ideal. If you have to work in noisy environments, **earplugs, earmuffs, or noise-canceling headphones can be invaluable**. The nice thing about headphones is that they also send a "do not disturb" signal to others.

Some of the worst distractions are notifications from your computer and phone, especially because you can become compulsive about checking them. You can be pulled off-task even when you aren't cognitively stuck. One study showed that people on average checked for messages every 35 seconds when messenger apps were left open.[12] On a brighter note, however, business employees who had their access to nonessential websites blocked for a week reported both deeper focused immersion and higher productivity.[13]

Do a sweep through the notifications settings on your devices and disable audible, visible, and vibrating alerts. "Do not disturb" mode may help. Use the Pomodoro Technique to keep you away from the internet or other distractions, or install a website blocker. Don't pity yourself that you live in a modern social media era where it's harder to get away from distractions. Even back in the mid-1800s, legend has it that novelist Victor Hugo, author of *Les Misérables* and *The Hunchback of Notre-Dame,* had his servant lock him naked in his study with a pen and paper to keep

POPULAR WEBSITE BLOCKERS
- Freedom
- FocalFilter (Windows)
- SelfControl (Mac)
- StayFocusd (Chrome)

him from the distractions that beckoned. (You'd think his books would have been shorter as a result.) Distractions are always going to be around—it's our job to figure out our own best ways to combat them.

Create a Ready-to-Resume Plan When an Unavoidable Interruption Comes Up

If you are interrupted by something or someone unavoidable, take a few seconds to mentally note where you are in your current task, and how you'll return to that task. This can be as simple as noting that you were three-quarters of the way down the page you were reading, and that's where you'll return your gaze when you finish the interruption.

This ready-to-resume plan reduces attention residue that can disrupt the new task. How? It provides the closure the brain is looking for—even if the closure is only temporary. A sense of temporary completeness allows you to fully engage in the interrupting task, even while it allows you to return to the original task more easily later.[14]

Take Frequent Brief Breaks

We already mentioned that the mental break part of the Pomodoro Technique is critically important. Too prolonged a focus doesn't give your brain time to offload the new material you're learning into long-term memory.[15] Your studying becomes less effective. In addition, specific areas of the brain can tire when you use them for a long time. Although researchers still don't know why, it's thought that just as muscles will tire from exercise, so can the brain tire from use, so-called "cognitive exhaustion."[16]

Short (5- to 10-minute) breaks involving complete mental relaxation—no internet, no texting, no reading, nothing at all—are the best for enhancing what you've just learned, because the new information can settle without interference.[17] This means

HOW MANY HOURS SHOULD YOU
STUDY PER DAY IN COLLEGE?

If you're a college student, we recommend that you study between 2 and 8 hours per weekday (in addition to classes), depending on your ambition and the rigor of your study program. The gold standard of study time is set by medical students with A grades. Above and beyond the hours they spend in classes, medical students generally study an average of 6 to 8 hours a day—studying more than 8 didn't improve grades. B and C medical students are more likely to study 3 to 5 hours a day.[18] The average engineering student studies about 3 hours a day, while social science and business students average 2 hours daily.[19] (Barb found herself studying engineering 6 to 8 hours a day, even though she generally took a slightly lighter than average course load. But this allowed her to earn A grades.)

you're not being lazy if you want to take a short nap or just do nothing—instead, you're being efficient.[20]

Breaks that involve something physical, like going for a walk or jog, or even just getting up for a cup of tea, are always a good idea. Part of the reason that breaks where you move around may be so valuable may simply be that you aren't thinking so much. Another reason is that movement and exercise themselves are helpful for the learning process—more on that later.

Music and Binaural Beats

Music seems to slow down learning for most students, especially in math.[21] You may feel better studying to music, and feel you can study longer. But that's because when you're listening to music, part of your attention is following the tune so you aren't working as hard as you could be. Music can also lead to multitasking as you switch between work and fiddling with your playlist. If you are getting good grades or evaluations for what you're learning, you're probably fine to listen to music. But if your feedback isn't what you'd like, or you're struggling to make headway with the material, we'd suggest backing away from music. There are hints, however, that those with attentional disorders may benefit from studying to music.[22]

Incidentally, there is a music-related phenomenon called "binaural beats." By wearing stereo headphones, the right and left ears can be supplied with two slightly different frequency tones—for example, 300 Hz and 320 Hz. Surprisingly, a person hears not only the original two tones, but also a third frequency—the difference between those two frequencies. In this case, the difference would be 20 Hz—called the "beat" frequency.

Researchers first became aware of binaural beats when they were investigating how the brain locates sounds.[23] Beginning in the 1970s, people began to explore possible changes in consciousness produced when the beats might shift, or entrain, brain activity toward the beat frequency. Most uses of binaural beats today are by regular people who download audio materials from various online sources to help them focus, remember, relax, or meditate. Since the beats can have a bland, monotonous sound, they are often embedded in music or pink noise.

You can explore studying to binaural beats, but be aware that the observed positive effects, at least in the baseline studies, are modest.*[24] And despite their claims, online sources for binaural beats can be of questionable legitimacy. Finally, research suggests that the effect of binaural beats on focus might be canceled by the effects of the music they might be embedded in.

Meditation and Yoga

Meditation has been suggested as a method for building focus. Overall, there are two general types of meditation—focused types, such as mantra meditation, and open monitoring types, such as

*To hear binaural beats, check out the samples at https://en.wikipedia.org/wiki/Beat_(acoustics).

mindfulness. Mantra-type meditation may provide more direct practice in building focus, although effects generally become apparent after weeks or months. Open monitoring–type meditation may help cognition indirectly, by improving mood. A challenge is that many past studies on meditation didn't follow proper scientific procedures, so more research is needed.[25]

There are some preliminary indications that yoga may have positive effects on cognition and may improve the connections of the diffuse mode.[26] (More on the diffuse mode in the next chapter.) But research on the effects of yoga is at an even more preliminary stage than meditation, so it's hard to draw firm conclusions.

• • •

In this chapter, we've covered how to focus on what you're learning. But sometimes focusing just isn't enough. What do you do when you get stuck? Read on!

KEY TAKEAWAYS
FROM THE CHAPTER

- **The Pomodoro Technique is one of the most powerful methods for tackling procrastination.** To do it:
  - Remove distractions.
  - Set a timer for 25 minutes.
  - Focus as intently as you can for those 25 minutes.
  - Reward yourself. Take a 5-minute or so mental break. (Use a timer for the break if necessary.)
  - Do another Pomodoro until the work—or you!—is done.

- **A pain in the brain can trigger procrastination.** Be aware of minor feelings of discomfort when you think about

something you don't like to do—these feelings can trigger
procrastination. That pain in the brain dissipates once you
get started.

- **In general, it's best to avoid multitasking.** But
  multitasking is not all bad—it can sometimes help you
  avoid becoming fixed on a cognitive approach that's going
  nowhere.

- **Set up a distraction-free environment.** Do a sweep
  through the notifications settings on your devices and disable
  audible, visible, and vibrating alerts. Keep your phone out of
  reach.

- **If you are called off-task by someone or something, try
  to take mental note of where you were so you can more
  easily return to it.**

- **Take frequent brief breaks.** If you go too long on any one
  task, you will inevitably tire.

- **If you like to listen to music when you're working,
  make sure it isn't distracting your attention.** Think twice
  about listening to music while studying unless you're already
  learning the material well.

2

How to Overcome Being Stuck

Olav once crashed a drone into the top of a tall tree, where it got stuck—too intertwined in the leafy branches to come free, far too high to reach by ladder, or even by throwing a rock. Climbing the tree wouldn't work either: the drone was caught in the thinnest branches. Olav felt as stuck as the drone itself. What should he do? Olav decided to do *nothing*. And that helped him get the drone down. How? We'll get to that in a moment.

Getting stuck and frustrated when you're learning is common— blank pages can leave you unable to think of a single sentence as you try to begin writing, or a new approach to coding can leave you stumped. We've mentioned a few tips previously to help you avoid becoming cognitively stuck, such as briefly switching tasks or taking brief mental breaks. But if you know a little about how your brain works, you can do even more to avoid this frustration and simultaneously speed up your learning.

The Focused *and* Diffuse Modes to Solve Problems Big and Small

The brain has two completely different modes of thinking and learning. The first is called **focused mode**. This was what the first

Focused   Diffuse

Learning involves going back and forth between
focused mode (left) and diffuse mode (right).

chapter of the book was about. It's exactly what it sounds like—you're in the focused mode when you're focusing on something. For example, you may be concentrating on an explanation of a physics problem. Or intently memorizing new vocabulary words.

The second mode is called the **diffuse mode**. This mode is also important for thinking and learning.[1] While you're in the diffuse mode, thoughts are still flowing through your mind, but you're not focusing on anything in particular. For example, you're in the diffuse mode when random thoughts pop up while you're standing in the shower, riding a bus, going for a walk, or falling asleep. When you're in this mode, your brain can connect different thoughts and ideas in a way that it can't while in the focused mode, which is busy suppressing everything except what you are focusing on. That's the reason that people get ideas and fresh insights when they go for a walk or take a shower.

Learning Something New and Difficult Means Alternating Between Focused and Diffuse Modes

Focused mode is all you need if what you're learning is relatively straightforward, perhaps related to ideas you've already mastered. For example, you're in focused mode when you're solving an uncomplicated addition problem, such as 14 + 32.

But what if you're trying to learn something new and more difficult? Let's say you're trying to understand the multi-pump system of the heart or the mathematical concept of a derivative, or master a physical skill such as how to do a double kickflip on a skateboard.* You might focus hard, then harder, and then even harder, and you still can't get it.

Strangely enough, allowing yourself to take a break, whether for several hours or overnight, often works magic. It's the magic of the diffuse mode. When you return your focus to the issue at hand, you'll have that "aha" insight that allows you to make progress on the issue you've been struggling with.

Let's use a metaphor for the focused and the diffuse mode to better understand the difference between them.

Think of your brain as a maze with learned concepts and procedures stored as pathways in different parts of the maze. When you're in the focused mode, your thoughts move along these pre-laid pathways.[2] You can see some of the pathways drawn into the maze on the next page.

When you concentrate on a familiar subject, like working a multiplication problem, you're using pathways you've already laid in one part of the maze. When you concentrate on another task, like conjugating a verb in another language you've learned, you use a pathway you've laid in a different part of the maze.

But if you're trying to solve a completely new problem—no pre-existing pathways—it can be tough for your focused mode, since there aren't any pathways for it to use.

So what, then, is diffuse mode? It's best to imagine the diffuse mode as a set of tiny drones that can zip quickly over parts of the maze. Because the drones can fly over the maze, they can easily make connections between parts of the brain that may not normally be connected. You use the diffuse mode whenever you're learning something new and difficult. The diffuse mode allows you to create the beginnings of a new neural pathway of ability and understanding.

---

*You form neural links in long-term memory when you're doing something physical as well as when you're doing something mental. More on that in chapter 6.

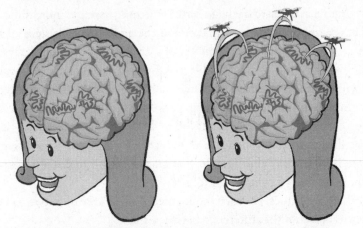

This picture of the brain as a maze gives you a metaphorical feel for how your brain works in focused and diffuse modes. The focused mode is when you focus on a preexisting pathway involving topics you already know. The flight paths of the little drones flying above the maze are your diffuse mode.

You're sometimes not aware of the diffuse mode when it is at work—but you become aware of that "aha" moment when you can suddenly understand something you've been struggling with. Like when you suddenly see how to solve a tough problem in data analysis, play a difficult passage on the guitar, or intuit a new marketing approach. The feeling of "aha" is when your tiny mental drone has suddenly made a new set of connections. What seemed confusing and incomprehensible before suddenly clicks.

Once the diffuse mode helps you make a new insight, the focused mode can build and strengthen the new ideas. This is why **learning often involves going back and forth between focused and diffuse modes**. You focus—meaning that you work intensely on the material—until you

KEY POINT
As long as you're focusing on a particular topic, you're *blocking* the activities of your diffuse mode on that topic. It's only when you get your focus *off* that topic that your diffuse mode can go to work on it.

But you must first focus hard on your learning challenge for the diffuse mode to then be able to work its magic.

begin to struggle. Then you take a break, and the diffuse mode works on the ideas in the background. Then you return to focus again, and it makes better sense. Back and forth it goes as your learning grows.

How to Get into Diffuse Mode

To get into diffuse mode regarding a topic, you must focus hard on the topic for a little bit or until you get stuck, but then *stop* focusing on it.

To then fall into the diffuse mode, it's best to do relatively mindless activities such as brushing your teeth, washing the dishes, ironing clothes, or as we'd mentioned earlier, walking, riding a bus, taking a shower, or lying down with your eyes closed. These are activities that may require a minor bit of focus (you don't want to walk into a wall), but not much. Your brain should be left free to wander.

When your focus tires, you naturally fall into the diffuse mode. How long you stay in diffuse mode varies. When you blink, for example, you're momentarily in the diffuse mode—unfortunately, this is too short a time to allow much mental processing to occur. When you daydream for a few minutes, you're in diffuse mode. You

---

EXAMPLES OF HOW TO EXPLOIT YOUR DIFFUSE MODE

- Start a difficult essay before dinner, so that diffuse mode can work in the background while you're eating.
- Begin a difficult problem set right before you take a break.
- Read difficult passages before going to bed—continue the next day.
- Rework an especially tricky or important problem before taking a shower.
- Review vocabulary lists right before going shopping.

can walk for hours in the diffuse mode. You naturally switch back and forth between focused and diffuse modes as the day goes by.

It should be clear now that falling out of focus into the diffuse mode doesn't necessarily mean you're wasting time. When it comes to learning, diffuse mode is your strategic weapon. Used wisely, your diffuse mode can give you valuable solutions and insights—this mode is strongly linked to creativity.[3]

Oddly enough, you can be in focused mode on *one* topic, and diffuse mode on *another,* which leads us to another powerful learning tool—the Hard Start Technique.

"Hard Start" for Homework and Tests

The Hard Start Technique takes advantage of your diffuse mode when solving difficult homework tasks or test questions. This technique is simple:

1. Scan over the test or homework problems and make a tiny check mark over any problem that seems especially hard.

2. Begin working on the *hardest* problem. You will probably get stuck after a few minutes.

3. As soon as you find yourself getting stuck, move to an easier problem.

4. Return to the hard problem later after you've done one or several easier ones.

> **KEY POINT**
> Learning when to *disconnect* from a problem, whether on a test or working on homework, can sometimes be as important as persistence. Students often lose points on tests because they get stuck and keep working fruitlessly on harder problems when there were easier problems they could have solved.

You will often be surprised that, when you return after working on another easier problem (or problems), you can make more progress on the hard problem. This occurs because, while your attention was focused on the easier problems, your diffuse mode was able to work behind the scenes on the harder problem.[4] Also, when you first started work on the problem, your tendency might have been to fixate on your approach, even if it was wrong. Temporarily switching what you're working on can allow you to reboot mentally, so when you return to the problem, you've got a fresher perspective.[5]

If you instead wait to do the hardest problems toward the end of the test or study session, you may be more mentally fatigued and unable to do your best. Worse yet, if you're trying to work the toughest problems at the end of a timed test, there won't be time for your diffuse mode to do any background processing.

The one caveat is that this technique only works if you've studied for the assignment or test—your diffuse drone needs preexisting chunks of knowledge or information so it can connect them together.

Hard Start also works well for essay questions. Start by drafting a structure or layout but without doing any actual writing. Then move on to some other questions. Return to the essay question after you've had time to do some diffuse mode background processing.

Use the Diffuse Mode When Writing First Drafts

One of the biggest challenges in writing a report or essay is to get the first rough draft done. This is because, when you're writing essays and reports, your tendency is to edit every sentence as soon as it comes out. Sometimes you'll reject a sentence while it's still in your head, before you even see what it looks like on the page. Nitpicking your work like this is like stopping to retie your shoe every time you take a step. You're not going to get anywhere fast.

The problem is that you're mixing up focused mode work (the editing) with more diffuse mode work (the writing). To stop yourself from doing this, cover up or turn off your screen and begin writing *without* seeing the words appearing on the screen. Or write by hand

WRITE OR DIE

Write or Die, an app by Dr. Wicked, allows you to set the number of words you want to write per minute and has various irritating options, including sounds and visuals, that kick in if you start falling behind your goal. There's even a "kamikaze mode," where your writing will unwrite itself if you're too slow. Oddly enough, the app can make writing fun.

without looking back at what you've written. Although this technique can feel odd at first, you'll soon get into the rhythm of getting words on the page *without* editing them. This allows you to make much faster progress in getting your first draft out. The whole point of a first draft, after all, is just getting it out on the page, no matter how bad you may think it is at the time. Later, you can edit.

Dividing your writing into diffuse (no editing allowed) and focused (editing) modes will do much to allow your writing to move forward more quickly.

Visit a Coffee Shop to Activate Your Diffuse Mode

When your study involves intense memorization, as with vocabulary or anatomy terms, it can be helpful to study in a quiet environment. But when your studies involve more conceptually difficult material—for example, trends in history, bridge construction, or grasping a difficult analytical concept—it can be better to study in an environment that occasionally disrupts your focus and allows your diffuse mode to pop up. This can allow you to gain new perspectives on these difficult-to-grasp ideas.[6] A coffee shop, with the hum of background conversations and occasional clanking of cups, provides many such triggers for the diffuse mode.

POPULAR "BACKGROUND" NOISE APPS AND SITES
- Coffitivity
- SimplyNoise
- Noisli
- myNoise

There are even coffee shop apps that play the sounds of a coffee shop so you can enjoy the ambiance no matter where you are.

How Olav Got the Drone Down

So, how did Olav dislodge the drone in the tree? He stopped focusing on how to get the drone down from the tree. This let his diffuse mode go to work. Suddenly, he was inspired to attach a fishing line to an arrow. He then shot over the branch where the drone was stuck. When he pulled on the line to shake the branch where the drone was stuck, it eventually fell.

• • •

In this chapter, we explored how you can couple mental focus with mental relaxation to solve difficult problems more easily. In our next chapter, we'll go deeper into the brain, to discover the best techniques for transferring information to long-term memory.

KEY TAKEAWAYS
FROM THE CHAPTER

- **Focused mode** helps you work through familiar problems, or to load difficult material into your brain so the diffuse mode can begin processing it.

- **Diffuse mode** helps you make sense of new challenges in whatever you're learning, whether it's understanding a new accounting concept, handling a challenging search engine optimization problem, or figuring out how to putt in windy conditions in golf.

- **Learning often involves moving back and forth between focused and diffuse modes.** It's typical to get stuck when learning something difficult—that just means it's time to move from focused to diffuse mode. Take a break, or work on something different, to let neural processing continue in the background.

- **Use the Hard Start Technique.** This means to begin with the *hardest* problems on tests or homework. Pause the hard problem when you get stuck so that you can work on another problem. Return to the hard problem after a while.

- **Do not edit when you're writing the first draft of a report or essay.** A good way to avoid editing is to cover or turn off your computer screen so you can't see what you're writing.

3

How to Learn Anything Deeply

Have you ever studied hard for an important test—reread notes, reviewed concepts, highlighted key parts—yet struggled on the exam?

Knowing how the brain learns will help you succeed with tests, and also help you build knowledge and skills that last. But learning doesn't take place in a vacuum. Physical exercise and sleep can allow your brain to more readily absorb what you're studying. In this chapter, we'll put all these ideas together. Let's go!

When You Learn, You Create Links

Whenever you learn something, you're simply connecting *neurons*—a basic "building block" cell in the brain. You have some 86 billion neurons—enough so you never need to worry about running out of capacity to learn.

The connection point between two neurons is called a synapse. Just as a person can reach out with an arm to touch a toe on the leg of the next person, the "arm" (axon) of one neuron reaches out to touch the "toe" (a dendritic spine) of the next neuron.

New knowledge takes form in your brain because you have created a new set of links among a little group of neurons in your long-term memory. This is true *whatever* you're learning—a new dance step, a new word in Latin, or a new mathematical concept.

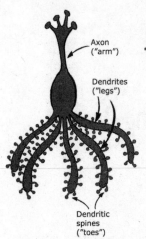

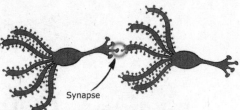

Axon
("arm")

Dendrites
("legs")

Synapse

Dendritic
spines
("toes")

Left: Neurons are the brain's building blocks.

Above: When you learn something, you form a link between neurons where the axon (the single arm) of one neuron snuggles up against the spine (a toe) of another neuron. This linking point is called a synapse.

When you think about something you have already learned, such as how to solve a problem like 4 × 25, or the Spanish word for "house," or the meaning of the word "condensation," signals travel between the neurons through the synapses you formed previously. Once you've learned something well, it's easier to think about it because you've formed strong connecting pathways between neurons in your long-term memory. The more and stronger the neural connections, the better you have learned something.

When you learn something simple, you create a short set of links. As your learning grows more complex, the sets of links become much longer and more entangled with other links. So if you learn to play a chord on the guitar, you can think of your small bit

The connected set of neurons here symbolizes something you've learned that's stored in your long-term memory. Connected sets of neurons can remind us of sets of links (placed behind the neurons).

The more you practice, the thicker and stronger the neural connections become. Learning more complex information well also creates longer sets of links—you can see how a small piece of information has only a short set of links, symbolized by the three links at the top, while more complicated information creates longer and longer sets of links. As you learn more, you also discover the connections and differences between different concepts—these are shown by the light sets of links connecting the main "concept" links. The more sets of links and connecting links that you create through understanding and practice, the more expert you become. Of course, in real life, learning involves many more neurons and links than we are showing here!

of learning as creating a little set of links. But when you learn to play an entire song, you create a much bigger set of links. Here's another example: when you learn the definition of the word "metaphor," you create a small set of links. But each time you see a new instance of a metaphor, you not only strengthen the set of links for the definition—you also make broader connections with other sets of links.

For any subject, skill, or discipline, having "preformed" sets of links, created through practice in a variety of situations, can be invaluable. For example, if you've created broad, rich sets of links, you can find it much easier to solve math problems in a variety of contexts. Or you can express thoughts in a new language, draw a picture of whatever you see, or easily draw up an algorithm that does what you need.

Study *Actively*, Not *Passively*

When you're learning, it's important to study *actively,* making your brain work hard or think hard. Don't just look at a problem solution. Instead, actively work the problem yourself. Or try to remember the main points from a video you just watched, or a book section you've just read. The mental effort you spend will help pull spines out toward the axons so that strong neural links can form.[1] This linking process continues during sleep. *Passive learning,* on the other hand, such as effortless listening or reading, isn't very effective. Your neurons just sit still instead of sprouting new connections that form new links. (In case you wondered, passive learning might be thought of as an inefficient focused mode—don't confuse it with being in diffuse mode.) **It's important to note that if you've done your previous studies in an active fashion, this also reduces test anxiety.**[2]

At the heart of active learning lies something called "retrieval practice." In other words, you want to see if you can pull information from your own memory, or work with it in your own mind, rather than simply looking at the material. The more you *retrieve* the material, and the broader the set of contexts you retrieve the materials in, the stronger and more broadly connected the neural links become.[3]

Oddly enough, the best way to put information *into* your long-term memory is to try to *retrieve* it from your own long-term memory instead of just looking at the answer.

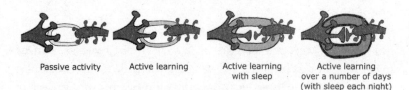

Passive activity   Active learning   Active learning   Active learning
                                                   with sleep   over a number of days
                                                                            (with sleep each night)

When you *passively* glance at material, as on the left, you aren't encouraging new neural connections to form. However, when you *actively* work with the material, as with the three sets of neurons shown on the right, you're forcing new spines to emerge and connect with axons.[4] The new strengthening connection is symbolized by the thickening and darkening link.

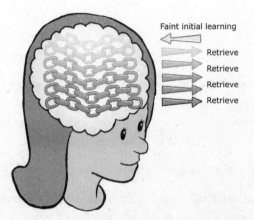

When you first learn something, it places a faint trace in your sets of links.
The more you retrieve those sets of links, the more you strengthen them.

There's another reason why retrieval practice is so important. When you try to retrieve something from memory, you get feedback that tells you what you know well (the stuff you're easily able to retrieve) and what you need to spend more time on learning (what you cannot retrieve). This feedback also helps you evaluate whether you've been spending your time productively or if changes are needed in the way you study. Retrieval practice is therefore also a *metacognitive strategy* that helps you evaluate your own learning. We'll talk more about the vital importance of metacognition and other metacognitive strategies in the final chapter of this book.

POPULAR DIGITAL FLASH CARD APPS
- Anki
- Quizlet
- Kahoot!
- GoConqr
- StudyStack
- Brainscape

General Techniques for Learning Actively

Here are some techniques that will help you approach your learning in an active way.

- Work out example problems *yourself,* without looking at the solutions. (If you have to peek partway through, after you finish the problem, go back and do the whole problem again.)

- Try to *recall* the key points from a book, article, or paper. Look away and see if you can recall the key idea or ideas. If what you're reading is difficult, it's best to pause and try to recall after each page of what you're reading.

- Formulate your own questions about the material.

- Take practice tests, preferably under time pressure that simulates the actual time constraints of the test.

- Find ways to re-explain key ideas from your notes or textbook in simpler terms, as if you're explaining them to a child.

- Work with others, either another person or a small group— meet and discuss the concepts, give mini-lectures, and compare approaches.

- Create flashcards, either by hand or using an app, such as Anki or Quizlet.

- Explain your thinking out loud to another person, or teach the key concepts to someone else.

- Ask a partner to quiz you. (The stress of being quizzed in front of a friend can mimic some of the stress of an actual quiz or test.)

- Take a practice test, even if you haven't studied much yet. (Guessing answers on practice questions has been shown to improve subsequent learning.)[5]

- Make your own practice test.

- Try to remember key points when you're doing mundane activities such as washing the dishes or going for a walk.

Challenge Yourself So You Can Advance More Quickly

Once you learn something, it feels good to practice it. It can feel so good that sometimes it tricks you. You can find yourself looking over older, easier stuff instead of digging deeper into the harder parts of the material. You can find yourself reviewing older vocabulary you've already learned in the new language you're expected to master, instead of tackling new words. Or you can find yourself practicing simple problems in data analysis instead of moving on to the more difficult materials, or singing a part of a song you already know instead of pushing on to the tougher part you haven't learned yet.

You might think that reviewing older and easier stuff prevents your neural connections from withering. But often, when you push yourself to go deeper into the material, you're building on what you already know. In other words, even while you're building new connections, you're still practicing with the older ones. **So if you want to advance quickly in your learning, you need to continue forming new connections in long-term memory, and not just reinforce the connections you've already made. This means it's important to keep pushing yourself every day with harder material.**[6]

How to Deepen Your Learning

To learn difficult concepts well, it isn't enough to simply recall dates, definitions, and facts. You'll need a deep understanding that allows you to explain, synthesize, analyze, and apply the concepts in novel situations. You cannot do this by simply memorizing (although as we discuss in chapter 5, *some* memorization can be helpful). To develop a profound understanding of what you're learning, it's important to actively connect what you're learning to other material you're learning or already know. Your neural sets of links should be connected to as many other sets of links as possible, to form a web of learning.

**Developing deep understanding means forming long links
that are also connected to other links.**

Elaboration

You can broaden your sets of links by actively thinking about what you're learning, as well as by writing and talking about it. A technique called self-explaining or elaboration can help.[7] With this technique, you actively try to explain, in your own words, what you're learning. When working on numerical problems, for example, stop at every step and ask the question "Why am I doing this step?" and then try to come up with an explanation. In one experiment, students who explained their steps when solving logical reasoning questions scored 90 percent on a later test. Students who didn't self-explain their steps scored only 23 percent.[8] For concepts that you're reading about, try to explain them as if you were teaching. Make your explanations different from the one you've just read. Try to simplify, improve, and provide examples.

Interleaving

Another important technique for broadening your sets of links is to interleave your studies. Interleaving helps you not only learn the concepts you're studying, but also understand the *differences* between them.[9] Interleaving means varying or mixing different concepts. It is the opposite of "blocked" practice, where you focus on plenty of

Interleaving gives you practice in choosing the right set of links.

practice with a single concept before dropping it to move on to the next concept.

For example, let's say you want to learn the styles of ten different artists. You could sit and view a dozen paintings by the first artist. Then you view a dozen paintings by the second artist, and so on.

But, tempting as it might be to study each artist in one blocked group, you should randomly *interleave* pictures by different artists. Look at a Manet, for example, then at a Van Gogh, then a Gauguin. This style of learning can seem more chaotic, but it gives you many more chances to contrast the differences *between* the styles, which helps you to much more rapidly develop your pattern recognition skills.[10] Work through the initial feelings of frustration, and you'll find yourself learning much faster.

It's similar in sports. Back in the old days, coaches used to expect their athletes to practice their skills in blocked format. In tennis, for example, forehand skills were practiced during one block of time, then backhands, then volleys. But researchers found that if players *interleaved* the different skills during their practice, randomly practicing forehand, backhand, and volley skills, players ultimately did better in competition. After all, part of playing tennis well means rapid pattern recognition: knowing which skill to employ as well as

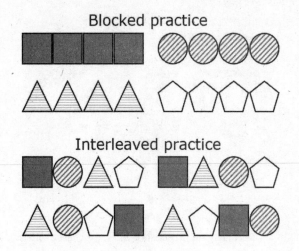

how to quickly alternate between those skills. Knowledge of inter-leaving has revolutionized some aspects of coaching.[11]

Unfortunately, many instructors and textbooks don't use inter-leaving. For example, when you're learning about different prob-ability distributions in statistics, your book might have a block of ten binomial distribution problems for you to practice, and then ten geometric distribution problems. You get practice with each distribution, but not with choosing between the two different dis-tributions. So it's often up to you to interleave the different topics and techniques. One approach is to make your own list of problems from different chapters to work on. Another approach is to take pictures of problems, artists, techniques, or what have you from different chapters and create flashcards, so you can practice looking at the problems and seeing if you know the proper technique to solve them.

Beware of Procrastination

You may think that procrastinating until right before a deadline creates stress, so you can focus better and learn more efficiently. Yes, such stress may help you complete straightforward tasks efficiently, but when it comes to learning, it can cause major problems. As we just saw, it takes brief periods of study each day over a number of days to grow a solid neural architecture of learning. If you procrastinate with something you're trying to learn, you're undercutting your ability to make progress. Give your brain the time it needs to learn a new subject. Remember, the Pomodoro Technique is golden!

Learning's Sweaty Little Secret: The Value of Physical Exercise

Researchers have long known that physical exercise helps us learn and form memories. Recently, researchers discovered a key reason why exercise is so helpful—it produces a chemical called BDNF (brain-derived neurotrophic factor) in the brain. BDNF is a protein that promotes sprouting of dendritic spines on neurons. Having spines available means it's easier to make new neural connections. Even a single exercise session can raise BDNF levels, but regular exercise raises the levels even more.[12]

Currently, there are no guidelines for exactly how much exercise you should do to benefit cognition, although researchers know that students who are physically active also perform better academically.[13] One study found that exercise using high-intensity interval training for 20 minutes, three times per week, over six weeks, resulted in a 10 percent improvement in memory of college students.[14] A meta-analysis found that even a single exercise session for

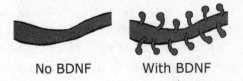

No BDNF   With BDNF

Exercise helps form a substance called BDNF in the brain. When you "sprinkle" BDNF on a neuron as shown above at right, dendritic spines emerge. BDNF is like a fertilizer for spines!

20 minutes can lead to immediate improvements in information processing, attention, and executive functions.[15] If the cognitive task is done right after exercise, this study found that it is better to do a light to moderate workout. But if there is a delay after the exercise, higher-intensity exercise is also beneficial.

The recommended guidelines for exercise in the United States are at least 30 minutes of moderate-intensity physical activity five days of the week, for a total of at least 150 minutes.[16] (Anything that gets your heart beating faster than usual counts as moderate.) Additionally, muscle-strengthening activities that involve all the major muscle groups should be performed on at least two days. Both types of exercise raise BDNF levels and make a myriad of other physiological changes that can assist with learning.[17]

A few very brief but intensive workouts during the day also help with fitness—and thus perhaps with cognition—for those who are restricted in both time and gym or outdoor access. One study found that warming up with a few jumping jacks, squats, and lunges and then climbing sixty steps (three flights of stairs) as quickly as possible, three times a day, increased aerobic fitness by 5 percent.[18] Incidentally, music can indeed make a grueling workout more pleasant.[19]

Cognitive Enhancements from Diets and Supplements

There's a natural tendency to want to try the latest faddish nutritional supplement, even if there's no research to support it. Ginseng and ginkgo biloba, for example, haven't panned out as effective cognitive enhancements.[20] But here are some dietary substances and practices that, as research has shown, have a small but positive effect on learning:

- **Caffeine (coffee, tea, guarana)** will act within 10 to 15 minutes to boost your attention.[21] The half-life of caffeine (the length of time it takes for half of what you took in to leave your body) is about five hours, so you can expect a bit of a lift during that time and even afterward, although people's metabolism varies. The drawback is that caffeine can inhibit sleep if you take too much close to bedtime, which in turn can inhibit learning.

- **Carbohydrates** such as those in a donut or a bit of sugar in your coffee can also make a minor boost in cognition within about 15 minutes after ingestion. This is because it provides glucose, a fuel used by the brain.[22] But be aware, too many carbs can make you sleepy. (Overeating in general disengages signaling pathways involved in cognition.[23])

- **Intermittent fasting** with two days a week of roughly 500 calories a day may also be useful in sharpening cognition.[24]

- **Flavonoids such as cocoa, green tea, and curry powder (curcumin)** can improve molecular architecture responsible for learning and memory.[25] But it can take up to six months before effects become noticeable.

Interestingly, **caffeine and carbs can combine synergistically** to have a more powerful impact on cognition than either substance alone.[26] Similarly, **a healthy diet, when coupled with exercise,** can improve cognition better than either diet or exercise alone.[27] Make sure your diet includes vegetables from the onion and cabbage family, a sprinkling of nuts, dark chocolate with little sugar, and fresh fruits of different colors.[28] For long-term physical health, processed foods are best avoided (so much for the donut before a test).[29]

Cognitive Enhancements from Drugs and Electrical or Magnetic Stimulation

A promising research area involves using drugs to get the brain to reopen the window of plasticity that made learning so easy when you were younger.[30] Certain drugs, it seems, can reset the brain so that it can temporarily become more plastic. But research in this area is in its infancy so this is very much a "do not try this at home" at present.

Some students might think that turning to synthetic cognitive stimulants such as amphetamines and modafinil can help, but they are often less beneficial than potential users think they are—and of course, they come with side effects and potent hazards related to addiction.[31]

Do-it-yourself noninvasive* electrical or magnetic brain stimulation might seem safe and helpful—if you were to believe companies that sell these products. But the truth is that the cognitive benefit (if any) of these devices is so small, and the set-up process so time-consuming, that even neuroscientists don't bother to use it on themselves. There are also important concerns about safety.[32] People who try it typically only use it for a little while before quitting.[33]

Surprise! Learning *Really* Happens When You Sleep

Learning means linking dendritic spines with adjacent axons to form neural connections in long-term memory. Interestingly, although a spine begins to bud when you start learning, that spine *really* connects with an axon when you're sleeping.[34] This is similar to how muscles grow during rest. And it's why you want to get a decent eight-hour sleep after a day of learning.

IDEAL SPACING INTERVALS
The golden rule of spacing is to wait until you've almost forgotten something before you return to the same topic.

This is also why it's so important to **space out your learning**. Ten hours of learning crammed into one day are not nearly as effective as ten hours of learning spaced out over ten days. This is because cramming forms initial connections but doesn't allow for them to be moved to long-term memory, where they can be strengthened and consolidated. The result? Those weak connections can quickly be forgotten. It's a lot like sports practice. No coach is going to say, "Let's practice ten hours today, and then meet this weekend for the tournament game!"

---

*The term "noninvasive" can be misleading. Although you are not sticking probes or surgical implements into the brain, you *are* putting electric or magnetic fields into the brain.

There's another reason sleep is important. During the day, your hard-at-work brain releases metabolites—toxic products. These build up as the day goes by, but they can't be washed away because your brain cells are like big boulders, blocking fluid flow. But when you go to sleep, your brain cells shrink. Voilà! Now cleansing fluids can flow through. Toxins get cleared out, which refreshes your brain for new learning.[35]

This also explains why naps appear to help with learning. One study of Singaporean students found that if they took a 1½-hour nap during the day, while sleeping 1½ hours less at night, they retained more and learned better in their afternoon studies.[36] Sleep researchers recommend eight hours of sleep (including the time it takes to actually fall asleep), so try to ensure you get that amount. (There is a special, rare "short sleep" gene—people with this gene tend to need only four to six hours of sleep at night. Odds are you do not have this gene—especially if you feel tired when you're running low on sleep.[37])

How to Fall Asleep More Easily

Long ago, during World War II, stressed-out pilots learned techniques to put themselves to sleep within about two minutes—here is a set of tips so you can do the same. First, as you're decompressing toward sleep, spend a couple of minutes writing a **to-do list** for the next day. This clears your mind. Also, avoid bright light from mobile phones, computers, or television screens before sleep by turning on night mode in your device settings.[38] If there is light in the room when you're sleeping, a sleep mask can be surprisingly useful. If possible, set the temperature of your room to around 65°F (18°C). Be sure to put your mobile phone in another room.[39] Remember that some physical exercise during the day—not right before you go to sleep—will allow you to better relax and sleep at

POPULAR TO-DO LIST APPS
- Todoist
- Trello
- Any.do

night. There's also good scientific evidence that weighted blankets can help you sleep better.[40]

Finally, once you're in your bed, follow the steps below.[41]

- **Start with thinking the word "calm."** This word will become your method of initiating the ability to relax.

- **Close your eyes and consciously relax all your muscles.** Start with your brow, which is often tense without you realizing it. The many muscles surrounding the eyes can also be tense too—relax them.

- **Breathe deeply and regularly.** Let your jaw, along with your tongue and lips, sag with complete relaxation. Try to breathe through your nose, keeping your mouth closed if at all possible. (The more you breathe through your nose, the easier you will find it to breathe through your nose. In other words, use it or lose it!)[42] Breathe in deeply, into your lower chest, for four seconds (that is, for a slow count of four). Then hold for four more seconds, and breathe out for six (this will really empty your lungs). Finally, hold for two seconds on empty. Then repeat the cycle, breathing in deeply for four seconds, and so on. This type of breathing balances both the oxygen and carbon dioxide levels in your body, and allows you to relax more deeply.

- **Your shoulders are often full of tension—let them go.** Let your chest feel like a relaxed jellyfish.

- **Tell each arm muscle—upper and lower, left, then right—to relax. Then your thigh and calf muscles in your legs—first left, then right.** Feel the relaxation happening.

- **Once you've relaxed your muscles, try to hold your focus on one mental object**—for example, a still cloud in the sky. Make sure you don't imagine yourself moving—in fact, the more you imagine yourself moving, the more it can wake you up. Alternatively, imagine a blank screen and let your imagination begin populating that screen with whatever dreamlike imagery begins to come to mind. You can nudge this a bit by consciously encouraging a positive scene of you in a safe place—a happy hero in your own phantasmagorical story. (But no thinking about your day job!)

- **Let your fears and worries go—mentally blank them out.** Some effective sleepers have a rule that they are not allowed to think between the hours of 10 p.m. (or whenever you go to bed) and 5:30 a.m., because they can't really solve life's problems before 5:30 anyway. If you wake up worrying about something and it's not time to wake up, remind yourself that you're not allowed to think about it yet and activate your "buzzer" to blank out your thoughts.

With practice (remember the importance of practice in learning!), you should also be able to use these steps to quickly fall asleep and stay asleep. If you wake up in the night, don't begin thinking. Just go back through your relaxation routine.

• • •

Learning anything deeply is a challenge—but part of the challenge involves making sense out of and retaining your learning. That's what we'll tackle in the next chapter.

KEY TAKEAWAYS
FROM THE CHAPTER

- **Learning means connecting neurons in your brain. To make learning deep and lasting, you have to make those connections strong.**
- *Actively* **engage with the material** to jump-start those neural connections, using **retrieval practice** whenever possible:
  - Work problems yourself—avoid looking at the solutions.

- Test yourself.

- Try to recall main ideas from a text.

- Explain key concepts in simple ways to yourself, or to someone else.

- Work with another person or small group that is as interested as you are in the material.

- Create study materials—flash cards, study guides, anything that requires you to process the information and put it into a new format.

- **Break up your learning into several shorter sessions over several days, instead of one ultralong session.**

- **Challenge yourself to progress faster.** When learning becomes easy, increase the level of difficulty.

- **To learn difficult concepts well, you need to actively connect what you're learning to other material you're learning or already know.** You can do this through *elaboration* or *interleaving*.

- **Don't procrastinate when you're trying to learn—it takes many days to build the solid neural architecture of good learning. The Pomodoro Technique can be very helpful here.**

- **Exercise regularly.** Exercise has been shown to help you more easily form neural connections.

- **Cautious use of cognitive enhancements such as coffee or tea, along with a healthy diet, can improve your ability to learn.**

- **Regularly get enough sleep.** Sleep is when your neural architecture grows. Spacing out your learning over several days will give more sleep periods, which will strengthen your learning.

4

How to Maximize Working Memory—and Take Better Notes

Genius mathematician John von Neumann was famous for performing extraordinarily complex mathematical operations in his head. Even at age six, he could do complex division with eight-digit numbers—once, when he spotted his mother staring aimlessly into space, he innocently asked her, "What are you calculating?"

Von Neumann could perform such remarkable feats in his head because he had an extraordinary working memory—the temporary mental space where you can hold and manipulate information.[*1] Working memory plays a key role in learning. It's what you use to solve problems and to make sense of material—an important prerequisite for storing that information in long-term memory.

Let's start this chapter with a quick overview of working memory. Then we'll move on to how your working memory impacts your ability to take in new information and take notes. Differences in your working memory capacity, as it turns out, can require different approaches to get the most out of your learning and note-taking.

---

*Short-term memory is usually thought of as involving the parts of the brain that can temporarily hold information in mind. For example, if you heard a person's name, that information would be stored (very temporarily!) in short-term memory. Working memory, on the other hand, generally includes short-term memory *and* your ability to manipulate—that is, do cognitive work on—the information. So if you associated the person's name, "Wanda," with a wand, and visualized a wand waving over Wanda's head, you'd be using working memory.

Working Memory—Master Trickster

You can think of working memory as an "attentional" octopus. When you focus on something, your octopus uses its arms to connect your thoughts. It can reach into long-term memory, for example, to connect sets of links there with the information that your eyes, ears, and other senses are currently taking in.

Normally, your working memory can hold about three or four thoughts or concepts—in other words, your attentional octopus has three or four arms.[2] But people can differ quite a bit in their working memory capacity—some can hold five or more pieces of information at once, while others can hold only two or three.

The arms are slippery—meaning the octopus can't hold on to thoughts for very long. For example, let's say you're taking a course in public speaking and are asked to give a two-minute impromptu speech about an embarrassing moment.

> **KEY POINT**
> The brain has two separate types of memory: *working memory*, which only holds information temporarily, and *long-term memory*, which is more permanent. True learning only takes place when information has been moved from working memory into long-term memory.

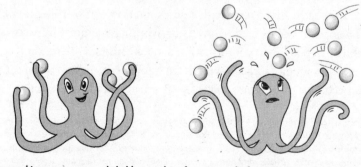

Your working memory can only hold up to about four pieces of information in mind at once. If it gets too many thoughts, it becomes overwhelmed. Information and thoughts can fall out.

Your working memory kicks in immediately. You hold the topic of your speech with one arm of your attentional octopus, silently repeating "embarrassing" to help you focus your thoughts. Your nervous awareness of the audience takes another arm of your attentional octopus. You send yet another octopus arm into long-term memory to search for embarrassing moments. It finds the memory of that time you spilled your drink all over your date on your first dinner together. That's a set of links that's easy to follow—you begin to talk about that night. Now you don't need to devote an arm to reminding yourself of the topic anymore, so instead you use that to touch on sets of links that you've practiced before to *look at the audience, smile,* and *gesticulate.*

Sets of Links in Long-Term Memory Enhance Your Working Memory

As we mentioned in chapter 3, whenever you learn anything—*really* learn it—you have created sets of links in your long-term memory.

The number of sets of links you can have in long-term memory is virtually limitless. But the number of arms of your attentional octopus—that is, what you can pull into your immediate attention—is quite limited. Sets of links serve as "preworked" extensions of the arms of your attentional octopus—these links boost the power of your working memory.

Your attentional octopus (working memory) can then grab these sets of links whenever you need to think about or use the ideas. Neural links that your working memory can access are what enable you to quickly create a pivot table in Excel, say a long sentence in Chinese, or solve that tough problem in circuit analysis, even though at first all of these activities might have seemed impossibly difficult.

Let's give you a trivially easy example. If you were given the letters "aueiutfbl," which you've probably never encountered before in that order, it would be hard to hold all the letters in your working memory. But if you're given the word "beautiful," which has the same letters, just rearranged into a form you've already learned, that word is easy to hold as one piece of information in your working memory.

In other words, with interconnected sets of well-learned links, you can effectively hold more in working memory at one time, since each one of the octopus's four arms can grab a piece of information that you've already learned.

Making the Best Use of Your Working Memory

If you struggle to understand the material, it's likely because your working memory is overwhelmed—it cannot handle all that difficult information at once. Here's what you should do:

Simplify

As you study books and articles, try to synthesize the key ideas, which are often surprisingly simple. Don't become overwhelmed with lesser details. If you are watching an instructor live or on a video, you might need to find ways to simplify what they are saying, as well, since even the experts can struggle with this. (Also remember that you can often find brilliant instructors on YouTube or platforms like edX or Coursera who are famous in part for their ability to make complex material seem easy.)

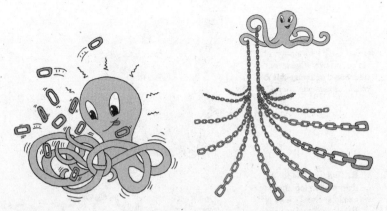

Your working memory may have to work hard to eventually create the set of links in long-term memory that can be easily accessed and used to solve problems and understand concepts.

Break Material into Chunks

Find a way to break your learning down into smaller chunks. Try to focus on the fundamentals.

Later, group the chunks of knowledge to form larger sets of links. For example, if you're trying to understand a difficult section in a physics textbook, pick the *easiest* example problem from the book and work it all the way through on your own, peeking at the solution only if you need to. Then work another example, and another, gradually moving toward more complex problems. This process will start placing foundational sets of links in your long-term memory—links that other links can attach to as you continue to master the material. Remember, if you get stuck, take a break or use your overnight rest to allow the diffuse mode to work in the background.

If you're overwhelmed in studying a language, try to focus on tiny bits at a time—a few words, which you can then put together into a sentence. When studying accounting, focus on understanding the profit and loss statement before you move to the balance sheet and cash flow statement. When playing a musical instrument, try

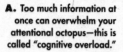

A. Too much information at once can overwhelm your attentional octopus—this is called "cognitive overload."

B. Find ways to put the information together in smaller chunks (sets of links). Your octopus can begin to gradually put those pieces together.

C. Eventually, you can hook the smaller sets of links together to form a large set of links that relates to the concept as a whole.

mastering individual short passages before linking them together to play the entire passage. And in karate, your teacher will have you practice with a sequence of smaller movements before you put them together into a seamless spinning kick.

Translate into More Understandable Terms

Another trick you can use to reduce the demand on working memory is to swap a fancy technical term for one that's easier to understand. For example, a "torque" might become a "twist." Or try to see and hear an explanation simultaneously. This makes it easier to understand, which is why videos are often easier to learn from than instruction manuals.[3]

In a greater sense, the challenge is to try to attach what you're learning to something you already know or are familiar with.

Make a Task List

When you're studying, try to clear your working memory of any thoughts that aren't directly related to the task at hand. Making a task list is a way to help with this. A task list allows you to transfer your thoughts from transient working memory to a safer place. Instead of trying to remember a dozen or more tasks, the only thing you need to remember is that you have a task list.

Put Something on Paper

Putting a key word, number, or formula on paper extends your working memory to the paper. Storing such information temporarily on paper instead of in working memory frees up capacity that can be used for other thoughts.

How to Take Better Notes

When you take in information from books, videos, or lectures, what you hear and see makes its way into working memory. But the information will disappear within seconds unless you deliberately focus on getting it into long-term memory. This is why note-taking is so powerful—it allows you to process the information, organize it, summarize it, and store it for further review and practice in order to create sets of links in long-term memory. Which brings us to an important question—what's the best way to take notes?

Prepare

If you're taking notes from a chapter or article, start by getting a rough idea of how the text is organized (As we discuss more deeply in chapter 9, even a minute or two of looking at bolded text and pictures with their captions can help.) If you're taking notes from a class lecture, you should read through or at least scan any associated material or assigned readings ahead of time. If you're watching a video, subtitles (closed captions) are sometimes available as a document you can glance through ahead of time. Such preparation will give you a structure that will help you make better and more organized notes.

Extract and Organize the Essence

Good note-taking—whether from books, videos, classes, or training sessions—demands that your brain be truly focused so you can extract the essence of what you're learning into your notes. The latest research says that typing and handwriting are each effective methods for note-taking.[4] So how do you take good notes? Here are two recommended ways:

- **Split notes**
  Before you start taking notes, make a vertical line one-third of the way across the page, as shown in the example. Then, try to capture the main ideas (not the word-for-word verbiage) in the space to the right of the vertical line. Use contractions and omit the tiny words your brain would fill in: the, a, like,

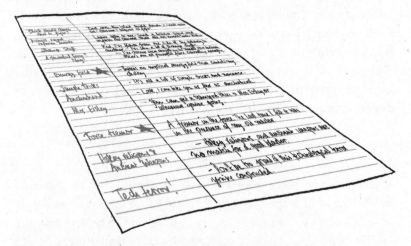

Example of split notes: Notes are written down in the right field and are later supplemented by summarizing key words or headings on the left.

thus. Use symbols (➜ + = ≠ # ✓ Δ), abbreviations (e.g., etc.)—whatever you can to make the task faster without losing material. Then, either as you're writing or when you're done, put summarizing words or brief phrases in the left-hand column.[5]

Later, when you review your notes, cover up the right-hand side and quiz yourself to see whether you can recall the deeper meaning from the summary words.

If something seems particularly important, or involves material that might be on the test, put a star next to it.

POPULAR NOTE-TAKING APPS

- Evernote
- OneNote
- Coggle (for mind mapping)
- SimpleMind (for mind mapping)
- Livescribe (allows users to replay portions of a recording made with a special pen and a paper notebook by tapping at the appropriate place on the notepaper)

- **Concept mapping**
  Concept mapping is an approach to organizing information to see how ideas and concepts relate to one another. It's used

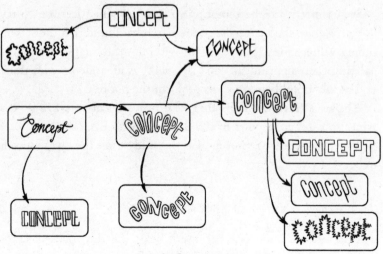

Example of a concept map.

for brainstorming (as when you're planning a project) and for taking notes. A concept map is created by writing down a few words with arrows between them to show relationships.

Review

The most important thing you can do with your notes is to *review them again before the end of the day*. Even if you're tired, spend a few minutes to pull the key ideas back into your mind. (Don't just skim over the notes.) A study of medical students found that "Students who made A's almost always reviewed the lecture the same day, while C students almost never did so.[6] A good review session including retrieval practice can be extraordinarily helpful in building those all-important sets of links. This review can actually be more valuable than the note-taking itself. Another study, for example, found the recall technique without any notes to be more effective for remembering and understanding the material than simply making a concept map without practicing recall.[7]

When you review your split notes, use the left-hand column material to quiz yourself on the more complete information of the right-hand column.

Sometimes it can be tempting to rewatch video lectures to try to better grasp the materials. But research has found that those who simply watch video lectures again without synthesizing and internalizing their main ideas don't do well.[8] And students who don't review any notes before exams perform the poorest.

Above all, remember that it's not the note-taking—it's what you get into your brain that matters. Even the best notes are worthless if they're not used as the study tools they were meant to be.

Partnering in Note-Taking

Taking notes is important because it helps you stay focused, especially if you're in a class or workshop. But for some, it can be difficult

MISSING CLASSES

If you must take a live class, and you have a long commute to get to that class, or you're taking a class where the instructor can't teach well, you might consider skipping class and instead watching lecture videos (if available), studying borrowed notes, and working problems yourself. Over the years, we've met many learners who have boosted their studying efficiency this way. Dr. David Handel, for example, who lived off campus and had a long commute to school, skipped many of his medical school classes, and used the extra three to six hours this gave him to learn more effectively.[9] He graduated number one in his medical school class. But do not try this unless:

- You are disciplined or motivated enough to self-study.

- You know how to self-study and what to do.

- You'll learn more by self-studying—either because you'll save a lot of time, or because the instructor is problematic.

- You won't miss any essential information—in other words, you can get the information you need from slides, friends, or written transcripts.

- You're not required to attend class, or penalized for your absence.

to follow an instructor while simultaneously taking notes because it may be too much for your working memory. In this case, a workable approach is to borrow notes from someone else or use shared docs to take notes collectively. Then you can help each other fill in the gaps. Research has shown that students who use other students' notes to review before exams do almost as well as students who have produced their own notes.[10]

• • •

In this chapter we've described how to maximize your working memory to make sense of and retain your learning. In the next, we'll go deeper into techniques to help you *remember* your learning.

KEY TAKEAWAYS
FROM THE CHAPTER

- **Working memory is only a temporary holding place for thoughts and information**. When you put something in working memory, it can seem as if you've truly learned it. But learning only happens when you secure the idea, concept, or technique in long-term memory.

- **You can typically hold perhaps four pieces of information in your mind at once**, because your working memory has only a few "arms."

- **You can compensate for the small size of your working memory by building sets of links in long-term memory.**

- **To make the best use of your limited working memory:**
  - Break material into chunks.
  - Translate what you're learning into more understandable terms.
  - Make a task list to clear your working memory.
  - Put ideas on paper to extend your working memory.

- **To take good notes:**
  - Use the split notes system or make a concept map.
  - Do a first-time review of your notes the same day you take them.
  - Make sure you actively practice with and try to recall the key ideas of the notes.

5

How to Memorize

How long does it take to memorize a shuffled deck of fifty-two cards? Alex Mullen from Mississippi, a medical student at the time, found a way to do it in less than 19 seconds. It's easy to think that Mullen is simply a genius, but as he himself says, his brain is nothing special. By using powerful memory techniques, anyone can learn to memorize information more easily and quickly.

Why Bother? The Value of Committing Information to Memory

In a world where you can look up any information, you might wonder: Is it necessary to memorize anything anymore? The answer is *yes*. You save time by knowing important information by heart. And during exams, interviews, and public interactions, you *can't* always just look stuff up. But perhaps the most important reason for thoughtful memorization is that it can help with complex problem-solving, and also help you acquire a deeper understanding.

Basically, whether you're grappling with the consequences of globalization or solving tough astrophysics problems, it's difficult to

KEY POINT
Memorizing key pieces of information releases mental power so you can understand more complex concepts and solve more advanced problems.

succeed without some memorization. What you need is quick and easy access to information from *inside* your head—that is, you want to have strong sets of links already established in your long-term memory about key information. Having these links readily available frees up working memory and enables higher-level thinking.[1]

Let's say, for example, you're asked to answer an exam question like "Compare and contrast the French and Russian revolutions." This question involves higher-order thinking—it's not just a trivial regurgitation of facts. But how could you even begin to formulate an answer if you haven't locked into long-term memory the key elements of both the French and the Russian revolutions, including such seemingly trivial concepts as the timeframes of the revolutions and the key concerns and unmet needs of the differing peoples?[2]

In chemistry, memorizing formulas for different acids will make it easier to make sense of information about those acids. And in physics, memorizing Bernoulli's or

> **KEY POINT**
> Not all memory techniques work for all kinds of material. When you need to memorize something, quickly consider different memory techniques and pick the one most suited to what you're trying to remember.

Poisson's equations supports your understanding of the relationships those equations represent. This "memory-understanding" relationship goes both ways: **It's easier to memorize information when you understand it well, but it's also easier to understand information that you have memorized.**[3,4]

How to Memorize Information with Memory Tricks

The most inefficient way to memorize is to simply look at information over and over again. You should instead, at the very least, use the recall technique—trying to retrieve information from memory. Of course, you would do this several times spaced out over a number of days.

But often, there's an even smarter way. Memory tricks—also

called mnemonics—can speed things up. You'll still need to practice and repeat to some extent if you use a trick, but much less than without it. **Memories made with memory tricks are not only easier to make, but "stickier." That is, they stay in long-term memory and can be more easily drawn from there into your conscious working memory when you need them.**

Verbal Memory Tricks

Here are some tried-and-true verbal memory tricks:

Acronyms

If you sprain your ankle, there's a good chance your doctor will recommend "Rest, Ice, Compression, and Elevation" and tell you to think of the word "RICE" to help you remember these four remedies.

You can use this trick with virtually any list of items you need to remember. Simply take the first letter of each word that you need to memorize and play around with the order to see if you can create a word. The names of the three major ancient Greek philosophers, for example—Socrates, Plato, and Aristotle—form the acronym "SPA."

Sentences

If your words don't form a neat acronym, you can create a quirky sentence in which each word corresponds to a word on your list. The sentence "My Very Elderly Mother Just Served Us Noodles" could represent the planets in order: Mercury, Venus, Earth, Mars, Jupiter, Saturn, Uranus, Neptune. Sentences can also help you memorize numbers. "How I wish I could calculate pi" helps you remember seven digits of pi, since the number of letters in each word corresponds to the digits of pi (3.141592).

Visual Memory Tricks

Have you noticed that it is easier to remember someone's face than their name? Your brain is super visual—almost half of the human cortex is involved in visual processing, while less than 10 percent is auditory.[5] This means your brain has a great visual memory. In one University of Iowa study, a group of people was shown a series of 2,560 images. A few days later, some of these pictures were shown again, but mixed with many images that hadn't been shown before. On average, people identified 90 percent of the original 2,560 images they'd seen.[6]

Vivid Image

The easiest visual memory trick is to create an image that represents the concept you're trying to remember. The more crazy, fun, and vivid your image is, the more it will stick in memory.[7] It's also good to build in some movement—as if the images make up a short movie clip. Say you want to remember that the moon landing happened in 1969. You could visualize a moon looking like a yin-yang (which in turn looks like the number 69). Vivid images can sometimes be strengthened by adding sounds or feelings. You could, for example imagine the moon swirling around with a swooshing sound.

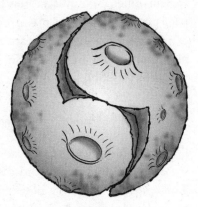

Memory Palace

When medical student Alex Mullen from Mississippi memorized an entire deck of cards in less than 19 seconds—setting a world record in the process—he used an advanced version of one of the oldest and most well-known visual memory techniques: the Memory Palace.

The Memory Palace, also called "the Method of Loci," is a technique in which you create an image for every concept you want to remember, and then anchor these images to familiar physical locations, often inside a building—hence the name Memory Palace.

Let's say you want to memorize the first five elements of the periodic table. With the Memory Palace Technique, the first step would be to come up with an image for each of the five elements. Here's one way to do it:

- Hydrogen: fire hydrant

- Helium: (helium) balloon

- Lithium: (lithium) battery

- Beryllium: strawberry

- Boron: boar

As you can see, for some elements, we have chosen images that are typically associated with the element, such as "battery" for lithium. For others, we have chosen images where the sound of the word leaves a helpful clue, such as "strawberry" for beryllium.

The second step is to place those images in a familiar location, such as your office or home, your friend's apartment, the street you live on, or your favorite park. We'll create a walk through your home for this example. Start outside the entrance door to your home where we've placed the fire hydrant (representing hydrogen)—it's leaking water and making a big mess. Next, you walk into the kitchen where you find a helium balloon (representing helium) hanging below the ceiling. You walk into the living room where a battery (lithium) is lying on the living room table. From there, you walk into the bathroom that is covered with strawberries (beryllium), squeezing the berries to jam as you walk around. Finally, you walk into your bedroom where a boar (boron) is running wild, creating a mess.

Once you come up with a Memory Palace, you'll still need to practice it a few times by going through your palace and visualizing the items. It can be a bit difficult to come up with good images when you first try this technique—but as with everything else, you'll get better with practice.

Whatever method you use to memorize, remember to strengthen those new sets of links by actively testing yourself over several days.

Metaphors

When you take a phenomenon you already know and use it to explain or understand a new concept, you're using a metaphor.* We've used a number of metaphors in this book, such as "sets of links" to explain how neurons form strong connections as you learn; or a "drone" to explain the diffuse mode.

A metaphor is a terrific way to simplify and grasp the essence of a topic. For example, programmers speak of stacks, queues, or trees. In biology, mitochondria are thought to be like batteries. And great literature is riddled with metaphors.

No metaphor is perfect, and you'll always find some parts of your metaphor that don't apply to your concept. For example, a drone will need to be charged regularly to continue flying, but the diffuse mode doesn't have the same need to be charged (well, unless you

---

*Technically, a metaphor is slightly different from a simile or analogy, but we'll just use the term "metaphor" in this book.

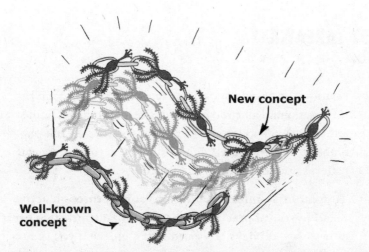

New concept

Well-known concept

Metaphors take sets of links you've developed for one concept, like a sponge, and use them to allow you to begin more rapidly building a set of links for the new concept, like wetlands. This approach relates to a concept in neuroscience known as "neural reuse theory," where sets of links developed for one concept can be reused to enable you to understand another.[8]

count sleep!). The point isn't to find the perfect metaphor, but to find one that works for the key aspects of the concept you're trying to understand. In the drone case, we wanted to highlight that the diffuse mode can allow your thoughts to easily move from one place to another, just like a drone.

To think of a metaphor, simply ask yourself, "What does this remind me of or resemble?" Try describing the concept to a friend. It may take several failed attempts to come up with a good metaphor, but that's okay. The time you spend debugging possible metaphors is time spent thinking about difficult concepts that can allow you to understand them even better. Part of the value of a good metaphor is that it helps you to not only understand what a concept *is*—it also helps you to understand what a concept is *not*.

• • •

Memorization works great for information, but what if you have to learn to solve a numerical problem? What you need then isn't *memorization*—it's *internalization*. That's what we'll turn to next.

KEY TAKEAWAYS
FROM THE CHAPTER

- **Memory and understanding are connected.** Having key information memorized so you can readily call it to mind releases mental power and helps you work more easily at higher conceptual levels. It also allows you to form mental patterns involving the subject you're learning.

- **To memorize information, test yourself through active recall (retrieval practice), spacing out your practice over at least several days.** To speed things up, see if you can use memory tricks such as:
  - Acronyms
  - Sentences
  - Vivid images
  - Memory Palace

- **Metaphors are useful for allowing you to more quickly grasp new concepts.**

6

How to Gain Intuition and Think Fast

Can anything good come out of playing video games all day long? For Tyler "Ninja" Blevins, the answer is yes. Blevins is one of the world's most famous gamers. For him, gaming is a full-time job that reportedly earns him around $20 million a year from sponsors and followers. Blevins typically spends ten to twelve hours per day playing games, which he streams to his more than 20 million followers. So far in this book, we've mostly been focusing on how to learn information from books, videos, and lectures. But what about when you want to become an awesome gamer like Blevins? Or become an outstanding race car driver, translator, mathematician, writer, or musician? Although conscious understanding built on strong sets of neural links can help, it's still not enough to bring you to top levels of expertise.

In this chapter, we'll look at how you can enhance and broaden your learning by tapping into your brain's powerful, mysterious (you're mostly not even conscious of it!) *procedural learning system.*

Declarative Versus Procedural Learning

Here's a little surprise. Earlier, we described how, when you learn something, you're depositing sets of neural links in long-term memory. But we skipped over the fact that there are two *completely different* ways you can deposit those sets of links.[1] One way uses the relatively quick-to-learn *declarative* learning system. This system is closely integrated with working memory, so you're mostly conscious of what you're learning (in other words, you can "declare" it). Most of the memory techniques we've learned so far in this book involve declarative memory.

But there is a second, equally important learning system: the procedural system. When you deposit sets of links through the procedural system, those links go into a different place in long-term memory than where you deposit links declaratively. It's like having two legs to stand on instead of just one. If you learn something using your declarative system, and then build on it by also using your procedural system (or vice versa), you have a much richer and deeper set of links for whatever you've learned.

Let's say you're learning how to serve a ball in tennis. In the beginning, you proceed consciously and declaratively by keeping your eye on the ball. When you first hit the ball, using your working memory to retrieve from declarative links the simple steps for hitting the ball, your serves are pretty bad.

But after lots of practice, serving the ball becomes an automatic procedure. This is because practice allowed you to place and strengthen "how to serve a tennis ball" links in the *procedural* areas of your long-term memory. When you draw on strong procedural links, your serve becomes smooth and natural.

For any given serve, once you decide consciously and declaratively (with the starting nudge of working memory), to serve a tennis ball, you begin to move your arm. The procedural system kicks in, and off the ball goes. Together, the declarative and procedural systems work hand in hand to give you a great serve.

For many decades, researchers thought that procedural learning only involved motor skills—like serving that tennis ball, kicking a football, playing the piano, or learning to type. Then researchers realized that the procedural system was also involved in habits, like putting your pants on before you put on your shirt in the morning, or nodding hello as you greet someone.

But now, researchers are discovering that the procedural and declarative systems work together in most kinds of learning, including writing, language, math, music, and, of course, video games. Unlike declarative learning, you're largely not aware of when you are learning through or using your procedural system.*

In essence, the procedural system is a black box—you can't see what's going on inside. The procedural system can receive input from your working memory (such as when you decide to try to serve a tennis ball), or input from what you sense (such as seeing a pothole ahead when riding a bicycle). But you are unaware of how the black box of the procedural system does its learning. All you know is whether you hit the baseball and it went to where you wanted it to go, or whether your bicycle glided smoothly around the pothole—or it crashed.

Despite its "black box" behavior, the procedural system can be extremely sophisticated—it can help you learn both simple and complex patterns without consciously thinking about them. Toddlers and children, with their exceptionally powerful procedural systems (their declarative systems don't come online until they grow older), pick up virtually their entire native language using their procedural system. Do you know how to solve a Rubik's Cube quickly? You're using your procedural system. Do you like to watch real-life surgeons in action on television? Their procedural systems play an important role in their expertise.

Well-developed links in your procedural system allow you to be lightning quick, even in stressful situations. These links involve not only actions such as serving a tennis ball, but also the ability to quickly see relationships such as 7 and 5 sum to 12 without having to think about them, or that a certain type of calculus problem requires taking a derivative. When you face a native speaker of the language you're studying, procedural links are also what enable you to speak comfortably and fluently, instead of finding yourself

---

*This is why, once you've driven to your house many times, you can become lost in thought and drive home on automatic pilot, without even being aware of how you got home. You first learned how to get home declaratively, but once you practiced enough, your procedural system can take over and you drive home unaware of your actions.

searching for words. Together, the speed and smooth confidence of the procedural system, coupled with the flexibility of the declarative system, can make you a learning force to be reckoned with.

Learn Using Both Systems

The declarative system is kind of humdrum. It learns best through step-by-step explanations. But the procedural system is unusual. It learns by intuitively feeling out patterns. In fact, you often cannot explain what you've learned procedurally, or at least not very easily: try explaining how to tie shoelaces without resorting to "watch me."

Let's cut to a key point. It's not possible to flip a switch and consciously move back and forth between declarative and procedural learning. As we mentioned, the procedural system learns by developing a feel for patterns. That happens through practice—and lots of it, in a variety of situations. So let's look at several different subject areas to get a feel for how to best use practice to improve both your procedural *and* your declarative learning.

Use Your Procedural System to Improve Your Problem-Solving Intuition, Speed, and Confidence in Math and Science

It's relatively straightforward to learn math declaratively. Just follow the step-by-step procedure you've been taught to solve the problem, and you're done. But this straightforward approach does little to help build links in long-term memory through your *procedural* system. A much better approach is to internalize key, exemplar problems. This helps develop your intuitive, fast procedural system.

To internalize a problem, you should pick a problem where the complete worked solution (not just the numerical answer) is available— *but don't look at the solution or procedure explanations.* Listen to your

internal voice instead—can you feel or intuit a whisper of your first step? (Remember,
your procedural system often can't put what
you're doing into words, as the declarative
system can. But it can hint!)

If you do get an intuition about what
that first step is, great! Do it. If you don't
get word from your intuition after trying
hard, take a peek, and then do the first step.
Then try to do the next step on your own.
And the next—all the way to the end of
the problem. Only peek if you need to—
and, of course, after you're done, to check
that you've done the problem correctly.

If the material is difficult, you may find
yourself taking a peek at virtually every

> **KEY POINT**
> The way to develop
> your problem-solving
> intuition is to try to
> draw the solution
> from *inside yourself*
> when working
> problems. Only peek
> at the next step of
> the solution if you
> absolutely have to,
> and then make sure
> you practice the
> problem again later
> on to make sure you
> can do it all without
> peeking.

step the first time you try to solve a problem—that's okay. Be sure
to work the entire problem by writing it down—don't skip steps.
Then try to work the problem again—hopefully without peeking
until you are finished.

Here's an example, just to give you a sense of what we mean.

Problem-solving steps	Examples of what you might be thinking as you look at each step. (As you advance further in math, you may not be able to verbalize the steps.)
$3(3 + x) = 21 + x$	How can I simplify this? Oh yes—multiply the 3.
$(3 \times 3) + (3 \times x) = 21 + x$	Okay, just need to multiply the items inside the parentheses now.
$9 + 3x = 21 + x$	Hmm—how do I get the x's by themselves more? If I move the 9 right and single x left, that'll do the trick.
$3x - x = 21 - 9$	At this point, I just need to do the simple subtraction.

2x = 12	To get x entirely by itself, all I need to do is divide both sides by 2.
x = 12 / 2	This is simple math.
x = 6	I'm done!

Most students do not do this extra internalization practice, and it's a big mistake that differentiates pro learners from ordinary learners.[2] Once you've internalized the problem you've selected, and several other problems that share resemblances—and differences—with the first, your brain begins to develop an intuition for how to solve these kinds of problems.[3] That's your procedural system in action!

In other words, as your brain internalizes seemingly simple but important procedures like "get rid of the parentheses" and "group the x variables on one side and numbers on the other," you begin to develop a deeper sense of the patterns involved in this and related types of problem-solving. This deeper, broader pattern sense can allow you to tackle problems even if the problems might seem superficially quite different from anything you've solved before.

This means, to develop your problem-solving intuition, you should internalize different types of problems, each over several days, until the solutions flow out easily with no peeking. (You don't need to wait to internalize one problem completely before you begin internalizing others.) Eventually, you should be able to just look at a given problem and step quickly through the various parts of the solution procedure in your mind, almost as if it were a song.

Don't be surprised if, the first day you try to internalize a problem, you feel it's just too hard. When you try again the next day, you'll be surprised at how much easier it is. And by the third day, it will start feeling natural—even, yes, intuitive!

You can quickly check your answers in math using a website like WolframAlpha or Mathway. Just enter what you want to calculate or know about, and AI technology will get you to the solution.

Remember, though, that you *can't* always look it up—you need a solid internal structure so you have an intuitive understanding of what's behind the computations.

Start Interleaving

The next strategy to use with your internalization technique is *interleaving*—that approach where you alternate different types of problems—for example, alternating module 3 problems with module 7 problems.[4] If you build interleaving practice into your internalization studies, you're laying a solid neural foundation. This will help you create sets of links about how to use specific techniques. It will also show you how those links relate to other sets of links involved in other techniques. You're allowing your procedural system to do one of the things it does best—detect different patterns. Later, on tests, you will find you have a natural, faster intuition about which technique or concept to bring into play.

What Should You Internalize?

How do you know what material is best to internalize? A great place to start is with the example problems that are worked out step by step in a textbook. They may seem easy, but they are often trickier than they first appear, and they usually demonstrate important concepts. Problems your instructor has worked out, as well as practice questions from old tests, are also great to internalize— that is, if you know that the solutions are correct. (As we mentioned earlier, taking practice tests is a great way to prepare for tests.[5]) The broader your pool of internalized problems, the easier you will find it to see analogies and transfer your skills to other, more distantly related areas.[6]

Use Your Procedural System to Improve Your Foreign-Language Studies

As a child, you naturally used your procedural system to learn to speak your native language. But as you grew older, you began to rely more on your flexible, fast-learning declarative learning system.

When it comes to learning a second language, however, that reliance on declarative learning is both good and bad.[7]

The declarative system can allow you to easily learn new vocabulary words and to quickly learn the patterns for verb conjugations or noun declensions. The problem is, when you find yourself in front of a native speaker, you often struggle. This can happen because you have put the links of learning in your slower-to-retrieve-from declarative memory. You haven't yet deposited sets of links in your *procedural system*. And it's the procedural links that give you easy, natural fluency in a language. The more you're able to develop your procedural links when learning a specific language, the more fluent your language skills become.

The Value of Retrieval Practice, Spaced Repetition, and Interleaving

No surprise, language learning needs a lot of retrieval practice—additional repetitions of retrieval practice almost always improve your ability to learn and retain the material. But realistically, there are only so many hours in a day and you're probably facing a deluge of new vocabulary words. So what's the optimal time for you to space out your repetitions? Minutes, days, weeks, or months?

Probably the most important question to ask yourself is how long you want to remember the material. If you have a test in a week, repeat the material each day for a week. If you want to remember it for a year, review it once every three weeks.[8] Once you're comfortable that you've got a word or phrase down, put the next repetition further out—expand the spacing gap. And remember that sleep and brief periods of mental relaxation help.[9]

Of course, interleaving and spaced repetition are also important—they improve not only declarative, but also procedural learning. When it comes to language learning, interleaving means *mix it up*! Don't just study static lists of vocabulary words written on a page. Instead, create either electronic or handwritten flashcards and shuffle them, so they're always in a different order. If you're studying three verb tenses over three weeks, don't just study one verb tense, and then the next and the next (textbooks and teachers love to do it this way). Instead, give yourself a chance to roughly grasp each tense as it is presented. Then, as soon as possible, start mixing up the tenses. This approach is harder and more confusing at first, but you'll learn better.

The best form of interleaving is to speak with a native speaker. You never know what words and sentences are going to be lobbed at you, or in what order. Start practicing with a native speaker as early and as much as possible in your studies, mistakes be damned. Get onto a website like italki and exchange conversations for free or hire a tutor. Remember that classroom instruction tends to emphasize declarative learning, whereas immersion (or as near as you can get to immersion) will help build your procedural system.

Sometimes people feel that they simply aren't capable of learning a new language. Often, they really could learn the language—it's just that the declarative approaches used in classrooms don't necessarily work well for those who tend to lean on their procedural learning system. For these more procedural types, diving into the deep end of the pool and speaking with native speakers from the very beginning, going back to look things up in books when necessary, can be a good approach.

GOOD LANGUAGE-LEARNING SOURCES

- Duolingo (an app that teaches you vocabulary, phrases, and sentences)
- Busuu (another language learning app)
- italki (a video chat platform that connects you to native speakers)
- Preply (a video chat platform similar to italki)
- Yabla (a platform with captioned videos that allows you to slow the speaking speed and break the video into small, easily repeated sections)
- FluentU (a platform much like Yabla, with a somewhat different selection of languages)

Use Gestures to Help You Remember Words

Intriguing research has shown that making a meaningful gesture while learning a new word in a foreign language can help you remember and understand that word better.[10] For example, if you're learning the word for "write" in another language, you might say "write" in the new language as your hands make a writing motion. If you're learning the word for "high," you might move your right hand up above your head. As you "drink," you would move your hand as if drinking from a cup. These gestures seem to help bind the sound of the word to its actual meaning.

Improving Your Writing and Artistic Skills

Famous statesman Benjamin Franklin used a special technique to improve his writing. Is it a declarative or procedural technique? We think it's a bit of both. In any case, it works.

To use the "Franklin approach," find writing you admire. Take a paragraph and jot a word or two down to remind yourself of the key ideas of each sentence. Then use those key words as hints to see if you can re-create the sentence. Check your sentence against the original and see which one is better. Does the original have better vocabulary? Better prose? If it does, you've just learned how you can improve your writing. Notice—you're not just memorizing other people's writing. You're actively beginning to build your *own* sets of links about how to write well. Eventually, using this technique, you will discover ways you can write better than the originals you're learning from.

Of course, this technique can work not only for writing, but also for art, language study, and other creative endeavors.

• • •

Knowing the best way to learn something will help you to learn it better and more quickly. But what if you'd rather play Fortnite (like

Blevins) than completing your study assignment? To address this problem, we turn next to the critical role of self-discipline.

KEY TAKEAWAYS
FROM THE CHAPTER

- Your brain has two pathways to store information in long-term memory. The declarative pathway helps you to get started with learning difficult topics or skills. The procedural system helps you to handle those topics or skills more quickly and intuitively.

- **Practice, especially spaced repetition and interleaving, helps you develop procedural sets of links in long-term memory.**

- **Internalize the procedures for solving scientific or mathematical problems** to help develop your intuition. Use interleaving to ensure you understand the differences between different types of problems.

7

How to Exert Self-Discipline Even When You Don't Have Any

In October 1912, a would-be assassin fired at President Theodore Roosevelt while he was giving a speech in Milwaukee, Wisconsin. The bullet penetrated the president's chest, but luckily, a small booklet absorbed much of the impact—the bullet never reached the president's heart. What Roosevelt did next shocked the crowd. He ignored the bleeding and simply carried on with his speech for another ninety minutes. Only after he finished did he bother to see a doctor.

Teddy Roosevelt was known for having extraordinary willpower. As a young boy with poor health, Roosevelt decided to strengthen his physical capabilities by introducing a heavy exercise regime, including weight-lifting, boxing, and many other sports. When his father died, Teddy responded by studying even harder, eventually graduating magna cum laude from Harvard. He is said to have read about a book per day—even when he was president—in addition to writing more than thirty-five books himself and 150,000 letters. How did he manage to do all this?

Theodore Roosevelt believed that the key to accomplishments was self-discipline—and that self-discipline was more important than talent, education, and intellect. "With self-discipline, all things are possible," he once said. But what is self-discipline? Can you increase it? And what do you do if you don't have much of it?

The Self-Discipline Challenge

Self-discipline is simply the ability to control yourself so that you can fight off temptations and in-the-moment distractions to reach long-term goals. When you know you need to study for an important examination but you're tempted by family and friends to spend more time with them instead, self-discipline can save the day. That is, if you have it.

This ability to make small sacrifices for future benefits is indeed, as Teddy Roosevelt believed, an important trait. Studies have shown that people who exhibit self-discipline are happier, healthier, wealthier, get into less trouble, and get better grades.[1]

Few of us have as much self-discipline as we'd like. We procrastinate, make impulsive decisions, and fall prey to temptations, only to regret it all later. The sad truth is that self-discipline is a limited resource—there's no quick and easy way to increase it. But even so, there *are* helpful approaches. It turns out that **a good way to become more disciplined is to take measures that reduce the need for self-discipline in the first place**. It's a little like getting a vaccination to prevent disease instead of treating the disease.

In other words, the key to self-discipline is to find ways to achieve your goals *without* relying on self-discipline. Let's look at some ways to do that.

Make Tough Choices Easy

Make it as *easy as possible* to make the right choices. Say you want to go to the gym every Thursday night but always find it hard to get your things together and head out. Try packing your bag the day before so that it's ready for you to grab and go when Thursday night arrives.

Or say that every time you get home, you find it difficult to start your homework. Try this: At the end of a study session, prepare your desk for the next day. Make sure the desk is cleared with your book already opened on the right page, with your pen and anything else you need next to it.

Suppose you always struggle to wake up in the mornings. A good trick here is to put your alarm clock on the other side of the room (or in another room) making it impossible for you to turn it off without getting up. Or you can download an app that requires you to solve problems to turn it off, such as Mathe Alarm Clock or Alarmy.

Eliminate temptations and distractions. **Research has shown that learners who remove temptations from their surroundings are more successful than those who try to rely on their self-discipline.**[2] Practically speaking, for example, if your mobile phone is distracting your studies, put it in another room. And if you're always tempted by sweets when you go grocery shopping, limit visits to the supermarket to once a week—after you've already eaten.

> **KEY POINT**
> To succeed with little or no willpower, remove temptations, distractions, and obstacles. Set things up beforehand to make it as easy as possible to get started.
>   Pick *one* small habit you'd like to establish or change in order to accomplish one goal you're working toward. Then, go for it!

Changing Your Habits

Habits form about all kinds of things. For example, looking sideways before you cross the street feels so natural that you don't even think about it. In fact, that's what a habit is—it involves the creation of a set of neural links that you draw on without having to think about what you're doing. This "mental autopilot" mode is the power of habit—it saves us mental energy. And, incidentally, it's a gift from your procedural system!

Habits can be good or bad. We may come home from work or school and plop ourselves in front of the television. Or we may start our homework. It depends on what we're used to.

A good way to minimize your use of willpower is to tweak poor habits. How do you do that? First, figure out the trigger to one of your undesirable habits. Then find ways to either remove the trigger or change your reaction to it. For example, do you always overeat or turn to unhealthy foods when you're starved? Try a a small snack to avoid becoming too hungry.

Changing a habit takes effort—the habit formation period, which requires at least some self-discipline, may take around two months.[3] But **building effective habits will enable you to become more productive while preserving precious self-discipline**. Even adding just one habit can make a surprising difference. For example, when Barb is procrastinating instead of starting a task, she turns to her habit of doing a 25-minute Pomodoro. She doesn't think about how much she dislikes the task. Instead, she just gets started—and thinks about how much she'll enjoy the reward at the end of the Pomodoro.

Plan Your Goals and Identify Obstacles

In the 1990s, German psychologist Peter Gollwitzer tried to understand why people don't reach their goals. He found that **having a strong desire to reach a goal isn't enough. What's necessary is a plan for when, where, and how you will reach your goal—and for how you will respond to obstacles.**

In one research experiment, Gollwitzer and his colleagues found that students who planned when and where they would study spent 50 percent more time studying than those who did not.[4] Another experiment found that students who planned their response to obstacles to their studying completed 60 percent more practice questions for an important test than students who didn't.[5] In yet another experiment, researchers from Germany and the UK found that 91 percent of participants met their exercise goals—if they made a plan that included when and where to exercise.[6]

Let's say you know you need to spend the weekend doing last reviews for a major set of final exams on Monday and Tuesday. At the same time, there are activities that are likely to lure you away. Will your self-discipline hold? According to Gollwitzer's research, you'll be much more likely to succeed if you plan when, where, and how you'll work. For example, you can plan to study Saturday and Sunday from 10 a.m. to 6 p.m., tucked away in the corner of a library.

You'll also be much more likely to succeed if you think about how you'll respond to temptations. Let's say that a friend asks you to go to a party. You might imagine yourself answering, "I can't. I already have other plans." (Stick with a vague response. It makes it harder for people to try to dissuade you—which friends will often try to do—because without details they won't have anything concrete to argue against.) When you plan and practice responses to obstacles in advance, those responses become natural and easier to carry out when the real temptation presents itself later.

Don't Forget to Recharge

It's easy to fall into a groove where, no matter how much you work, it doesn't feel like it's enough. This is a rapid path to burnout. It's important to set aside time for having a life, spend time with the ones you love, and have fun. If you build your break and reward times into your schedule—for example, no work between 6 and 9 p.m. each evening—it can enable you to focus more fully when you're studying.

Sometimes, studies and work can be so demanding that it's difficult to take much time off. In that case, find something you really enjoy that you can look forward to. One medical school student we know, for example, overwhelmed by the firehose of information she had to learn, lived each week for the one hour of her favorite television show.

Involve Others

In writing this book, Olav regularly emailed Barb to tell her when she could expect to receive updated versions of the manuscript. Barb didn't ask for it, and there was no hurry. So why did Olav do it? He finds that when he promises someone that he'll complete a task within a few days, the sense of urgency allows him to get the work done.

When you have a task that requires self-discipline, see if you can create a meaningful deadline or commitment by involving others. Say you need to study on a Saturday but know it will be hard to do. Find someone else who's also planning to work on Saturday and agree to meet up and study together. This outside accountability will push you to get the work done because you want to keep your word and your appointment.

• • •

Let's return to President Roosevelt. Averaging about a book per day, he might have been the best-read man in the United States at the time.[7] But when it came to reading, Roosevelt didn't need self-discipline to get it done. Why? Roosevelt loved to read! You only truly need self-discipline to do the tasks that you're not that motivated to do. If you can increase your motivation for studying, that might be the best remedy of all. That's why we'll next turn our focus to the crucial role of motivation.

READING TIP

If you want to read books but find reading to be a challenge, just set yourself a goal of, say, twenty pages a day. By the end of the year, you'll have read over twenty books!

KEY TAKEAWAYS

FROM THE CHAPTER

Self-discipline is important for success but also a limited resource.

- **Find ways to overcome challenges without having to rely on self-discipline.**
  - **Remove temptations, distractions, and obstacles from your surroundings** and make it as easy as possible to make the right decisions.
  - **Change existing habits that may be harming your ability to study by finding the habit cue and changing your response to it.**

- Plan your goals and identify obstacles and the ideal way to respond to them ahead of time.

- Involve others in your work to increase your commitment.

8

How to Motivate Yourself

Tom Sawyer looked at the big fence in front of him. Next to him stood a bucket of whitewash, waiting for him to spread it out across the bare board planks. Tom sighed. Normally, he would be out on a small adventure by now. Instead, he had been sent to whitewash his aunt's fence as a punishment for skipping school to go swimming. He took a deep breath and started the boring job of stroking the board planks. Right, left, soak the brush in whitewash, and repeat. To make matters worse, kids passing by would stop and make fun of him.

But within minutes of the start of this famous story—*The Adventures of Tom Sawyer*, by Mark Twain, Tom had gotten all the other kids on the street to paint the fence for him. He had even convinced the kids to give him gifts to be allowed to paint. How did he motivate the other kids to do it?

It's About Effort

Say you want to do well and be successful in your studies. Does that mean you have a lot of motivation? Not necessarily. **Motivation isn't about how much you *want* something; it's about how much *effort* you're willing to exert to get it.**

What gives you this willingness to exert effort? Researchers believe that the chemical dopamine plays a key role. They found that rats would work much harder to obtain food if dopamine levels were high in brain regions related to our emotions.[1] Later experiments have shown that humans will also work harder when they feel the pull of dopamine.[2] So dopamine is like adding a turbocharger to an engine—it boosts your drive. And although you could insert an electrically charged wire into your brain to increase the flow of motivational dopamine (hey—it works on rats!), we'll look at some better and safer ways.

Motivation as a Sum of Parts

Let's revisit Tom Sawyer. After a few failed attempts, Tom managed to motivate the other kids on the street to paint the fence for him. We tend to think that motivation is something that we either "have" or "don't have." But as Tom discovered, **motivation can be created, strengthened, and maintained through intelligent tactics.**

So what does it take to motivate yourself? It's a little like asking what it takes to move a car forward. An engine, gas, oil, wheels—altogether, a lot. In the same way, there are many influences on motivation. In this chapter, we'll explore how finding value, experiencing mastery, setting goals, and working with other people can increase your energy and desire to work on difficult tasks.

TO BUILD AND MAINTAIN YOUR MOTIVATION

- Find value in your studies.
- Work to experience feelings of mastery.
- Set goals.
- Work with others.

Understanding these factors can allow you to create and sustain your motivation, even as times get tough.

Value—Find What's in It for You

We are motivated to spend time on activities that we enjoy or benefit from. If you're learning Spanish but don't enjoy it, don't find it useful, and don't care about a grade, you'll find it difficult to motivate yourself to study. Whether you find an activity enjoyable and beneficial depends a lot on your perception. That explains why two students can differ so much in their motivation, even when they are performing the same task.

Luckily, you can change your perception—as Olav discovered in high school when he took a part-time job at a busy restaurant washing dishes. Twice a week, he worked nonstop from 4 p.m. to 2 a.m. When he compared studying to hard labor in an overheated restaurant kitchen, studying wasn't bad at all.

Changing perceptions was also how Tom Sawyer succeeded in motivating others to paint the fence. Tom realized that people want what is difficult to get. So he reframed the painting task from a mundane job to a once-in-a-lifetime opportunity. When speaking to his friend Ben, Tom refused to call painting "work" and instead said he enjoyed it. When Ben asked if he could try to paint a little, Tom refused, saying that his aunt was picky about who painted the fence. Making painting a hard-to-get activity motivated Ben to offer his apple in return for some time with the brush. Tom agreed, and soon all the other kids were trading away their toys to be allowed to paint.

One way to reframe your perception of a task is to make a list of all the benefits you can think of. *Dinner with friends without a guilty conscience? Look good when you get together with your team?* If you take the time to compile reasons, ideas may emerge that you hadn't thought of before, and this can motivate you even further.[3] Your list can include how the task moves you closer to your goals. For example, if you want to earn a certificate in project management, you might think, "Doing this homework set will move me closer to that certificate. Not doing this homework set will move me away from that certificate."

Another way to make a task valuable is to reward yourself when you've completed some portion of the task. A reward can be simply taking a break where you listen to your favorite music, or having the chance to watch your favorite television show, to hang out with friends, or even just to go for a walk. (After sitting awhile, walks can feel good.) Rewards can be especially useful when you face hard or boring tasks, which is another reason why the Pomodoro Technique is so powerful. But be careful of using your mobile phone during your break—remember, research has shown that the focused attention you put into the phone means you're not actually getting the mental break you need.[4]

Mastery—Experience Progress

A feeling of mastery is a powerful motivator. As you work toward mastery, your studies should fall in the "doable with effort" range. You can grow frustrated if the task is too hard. But if things are too easy, you'll get bored.

If a task is too difficult, try to break it up, seek help, or find better resources. Find an online video on the topic or post your question in a discussion forum online. Another technique is the "rubber ducky" approach, where you place a rubber ducky, or whatever object you'd like to speak to, in front of you—you then try to explain out loud where you're stuck. You can also take a step back and restart at a lower level. For example, when Olav studied calculus at university, he hadn't studied math in two years and found himself struggling. So he dug out his old high school math book and spent a weekend reading explanations and redoing calculus problems. That gave him more-appropriate practice problems and explanations, which allowed him to begin to grapple successfully with university-level math.

Incidentally, getting help is sometimes the best approach. When Barb was preparing for a midterm for her first programming class in college, she made a list of a dozen questions. To get answers, she went to the tutoring center on campus. Later, she discovered that nearly every one of her questions was covered on the test. Taking the time to ask for help at the tutoring center made a big difference in her understanding of key aspects of the material—and her grade. This further increased her motivation.

How you deal with setbacks also matters. If you see setbacks as failures, it's easy to be demotivated. Instead, **you should consider setbacks as valuable opportunities to learn and grow.** Of course, when you're in the throes of what seems like failure, it's hard to be upbeat. But just keep in mind that what seems like a really big deal at a given moment probably isn't in the greater scheme of things. Or as Nobel Prize winner Daniel Kahneman said, "Nothing in life is as important as you think it is while you are thinking about it."

Goals—Aim for Something

When Olav was asked to start swimming to recover from an arm injury, he realized that he didn't enjoy swimming back and forth in a narrow lane. To make swimming more bearable, Olav discovered that setting a one-kilometer (0.6-mile) distance goal for each swim worked perfectly. He kept track of his distance, and every time he ticked off another kilometer, he felt great.

Setting goals is a great motivational tool. We recommend that you **set *long-term* goals, *milestone* goals, and *process* goals.**

Long-Term Goals

Your long-term goal should be something that excites and gives you a good feeling whenever you think about it, such as one day becoming a doctor, traveling the world, or starting your own business. **Keep a photograph or physical object within handy view that**

represents your long-term goal. That will act as a reminder that motivates you to persevere along the sometimes frustrating road to get there.

With long-term goals, you can also use a motivational technique called "mental contrasting." This means contrasting your life today with how your life will be once your long-term goal is achieved.[5] Say your long-term goal is to become a doctor. But right now, you're in college eating macaroni and cheese for dinner, working part-time at a job you don't enjoy much, and finding it hard to keep up with your studies. Now close your eyes and visualize how your life will be as a doctor. Imagine what a typical day will be like, where you will live, where you will work, and how your usual workday will unfold. This contrast between the future and today will energize you to keep working toward your goal.

You can also use negative contrasting as a motivator. For example, when Barb was trying to switch gears and learn engineering in her late twenties, it wasn't easy. Sometimes her resolution flagged. But then she would think back on how it felt to be working as a private in the army, where she didn't have much say about where her career went. She didn't want to be in that situation again. Engineering provided a good pathway out of the kinds of routine jobs Barb didn't want to be stuck in.

Milestone Goals

Your long-term goal should also be supplemented with some near-future milestone goals—**goals that act as steps on the road to reaching your long-term goal**. Such goals could be reaching a certain grade point average and goals for your individual subjects.

Process Goals

Milestone goals, in turn, should be supported by process goals—that is, **actionable goals that say how you are going to reach your milestone goals**. "Study math for one hour each day" and "Learn ten new words every day" are examples of process goals.

There is a tried-and-true piece of advice that **your long-term, milestone, and process goals will be more motivating if they are SMART: Specific, Measurable, Ambitious, Realistic, and Time-limited**.[6] In other words, you should be concrete about each goal and ensure that you can measure progress and attainment. Your goal should be a little difficult, but not unreachable, and you must be able to set a deadline for it. For example, the goal "to do well" is not a SMART goal because it's not specific, it's hard to measure, and it has no deadline. On the other hand, to "get an A on my next machine learning assignment" can be a SMART milestone goal, just as "to work on this assignment for 45 minutes a day for the next five days" can be a SMART process goal. Olav's goal of swimming one kilometer every week also meets the SMART criteria.

> **GOOD GOALS ARE "SMART"**
> **S**pecific
> **M**easurable
> **A**mbitious
> **R**ealistic
> **T**ime-limited

Find Someone to Work With

Watching a movie, going for a walk, and having a meal can all become more fulfilling if we share it with someone we like. That's because people have a built-in need to be socially connected and to secure the love and respect of others.[7] And face it—studying is more enjoyable if you have someone to study together with. If some of your friends are serious about their studies, it's even better. How to find such friends? If you're taking online classes, keep an eye on the discussion forums—you can meet very helpful people that way. If you're taking live classes, watch for students who ask good questions. Even if you're shy, it can be worthwhile to go up to that person, introduce yourself, and start a conversation, perhaps by asking a question you might have been too reserved to ask in front of the class.

If you're working on a difficult task, having a study buddy or a small study group can motivate you. The task doesn't change, but if you're figuring out concepts and problems with others, it can become as exciting as a soccer team working together to win a game. If your buddies are ambitious and driven, some of their motivation may even transfer to you—a phenomenon called *motivation contagion*.[8] Working with others can also deepen your learning—in part by having your friends point out where your thinking has gone wrong. Just ensure that if you meet to study together, you do just that. If you can't, it's better to study on your own and meet your friends for social activities later.

> **Forming a study group** is a great way to facilitate learning and stay motivated. Discussing the study material and hearing what others think can give insight into the key points of the material.

Let's return a final time to Tom Sawyer. Tom tested a number of motivational tactics, including begging for help and paying others, before he found one that worked. And in that lies another important point about motivation. Since many tactics can give rise to motivation, you may have to experiment a bit to find the best one for you.

• • •

In the next chapter, we'll move on to how to be more productive with one of the most common ways of studying: reading.

KEY TAKEAWAYS
FROM THE CHAPTER

- **Motivation is not something you either have or don't have**—it can be created, strengthened, and maintained using different techniques.

- **The best way to motivate yourself will vary,** depending on the cause of your low motivation. It's therefore wise to try different strategies.

- **Remind yourself of all the benefits of completing tasks.**

- **Reward yourself for completing difficult tasks.**

- **Make sure that a task's level of difficulty matches your skill set.** Ask for help, break a task up, or allow yourself more time (if you can!).

- **Use mental contrasting,** both positive and negative.

- **Make your goals and ensure that they are SMART (Specific, Measurable, Ambitious, Realistic, and Time-limited).**

- **Surround yourself with other students who are interested in the topics at hand.**

9

How to Read Effectively

In 2007, speed-reading champion Anne Jones sat down in a London bookstore to read *Harry Potter and the Deathly Hallows*. After forty-seven minutes, she had finished the entire 784-page book. At that rate, Jones was reading at 4,200 words per minute—approximately twenty times faster than the average person. If you could double or triple your reading speed—while still understanding what you read—it would be useful. But is it possible?

How to Actually Read Faster, and Why Speed-Reading Techniques Don't Work

To better understand what you can do to read effectively, it can help to first know a little about the reading process itself.

When you read a word—say, "car"—you first recognize the word. Next, you silently pronounce or "subvocalize" it before finally converting it to the idea or concept it represents.[1]

To recognize a word, the eye stops to focus on it for around 0.25 seconds. Then the eye jumps to the next word, where it stops and focuses again before jumping to the next word, and so on. These jumps take less than 0.1 seconds.

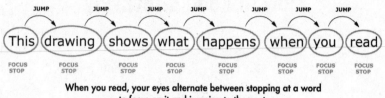

When you read, your eyes alternate between stopping at a word to focus on it and jumping to the next.

Some speed-reading programs claim to increase your reading speed by reducing these jumps and stops. Such programs will ask you to read three words at a time, or if you're using an app, they'll flash three words at a time, eliminating the need for the eye to jump altogether.

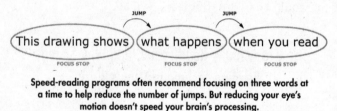

Speed-reading programs often recommend focusing on three words at a time to help reduce the number of jumps. But reducing your eye's motion doesn't speed your brain's processing.

However, research has revealed that your brain is processing the previous word during the jump to the next word.[2] *The jump is not what slows you down.* What's holding everything up is the mental processing—the recognition of the word, subvocalization, and the conversion to idea or concept.

So if you try to increase your reading speed by cutting back on focus stops and jumps, you're not addressing the real constraint on your reading speed. **If you do want to read faster, you need to speed up the word recognition and conversion to concept.** You can do this by having a good vocabulary, the right background knowledge, and plenty of experience with reading.

Effective Reading Is All About Comprehension

Even though you can increase your reading speed by building your vocabulary and background knowledge, there does seem to be a limit. Research shows that fewer than 1 percent of people can read at speeds faster than four hundred words per minute without loss of comprehension.

For most people on typical reading materials, a comfortable reading speed will be around one hundred to three hundred words per minute. But that's okay, because **effective reading isn't about speed; it's about understanding what you read and remembering it**. You may notice that if you read a lot of highly technical material, your general reading speed even on non-technical material may slow down. That's perfectly normal. Besides, slowing down a little can improve your comprehension.

So how did Anne Jones read the entire Harry Potter book at 4,200 words per minute? It's worth noting that Jones's comprehension was not scientifically measured. Instead, she simply gave a synopsis of the book to reporters. As reading experts argued, she could have done this because she had already read all the earlier Harry Potter books, and she might have had a lot of practice in summarizing books based on bits and pieces of information.[3] Could Jones have used the same techniques to carry her as swiftly through a textbook on vector calculus? We doubt it.

Preview the Material

It's much easier to put together a jigsaw puzzle if you've already seen a picture of what the completed puzzle should look like. In the same

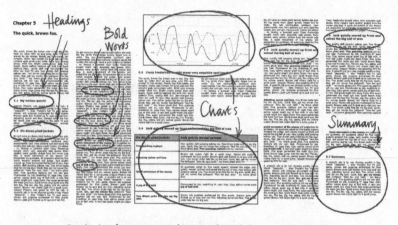

Previewing the text means skimming through the summary and the structure of the book to get a big-picture feel of the text.

way, a quick, "big picture" overview of a new chapter can give a useful sense of what's going on before you dive into the details.

Most textbooks are designed with a specific layout that includes the summary, chapter goals, end-of-chapter questions, bolded headings, pictures, and captions. If your book is like this, before you start reading thoroughly, take a few minutes to skim through. What's the big picture—where's the chapter going?

You may not feel that you're learning much from such a quick review, but the point is just to develop a framework that will help later, as you read through at your usual speed. If you find that you get lost in the details as you read, you can do another preview to remind yourself of the big picture.

Avoid Passive Reading Practices

Let's say you're determined to master a new chapter in your computer forensics textbook. You highlight part of what you're reading—this makes you feel like you're actively engaging with the text. Then you reread the chapter, your eyes skimming more easily over the material, and notice that the information now seems more familiar. Surely, you think, you must be burning the information into your brain.

But your intuition is deceiving. When you highlight material, for example, you activate the parts of your brain involved in the movement of your hand. That's why you feel you're doing something active. But the "move your hand" part of your brain is not necessarily where the learning links of conceptual understanding are formed. In other words, **highlighting or underlining is passive, at least from the perspective of learning. It is not doing much to create the neural links of learning in your long-term memory.**

Rereading the chapter right after you've first read it can provide a deceptive familiarity. The words become more familiar, but you're not truly grappling with the material—the so-called fluency fallacy. Your neurons aren't being forced to make connections. Along these lines, Carol Davis, an instructor of electrical engineering technology, notes, "Rereading takes away from time you could be actively learning and makes you feel like you've been studying forever without getting anywhere at all."[4]

If you want to reread, wait at least a day and remember to supplement it with more active learning methods. The exception to this is when the material is particularly dense, with many new and complex concepts. In that case, you may find yourself rereading paragraphs or sections repeatedly before you start understanding what's going on. Ultimately, your first read, followed by sleep time to help with consolidation, can give a sense of the big-picture concepts, while your second read can allow you to focus more on the details.

Practice "Recall Reading"

One of the best ways to read actively is to use a technique we've discussed earlier: *recall,* (retrieval practice) by researchers.[5] The recall technique allows you to not only remember, but also better understand key ideas. One study compared rereading to recall and found that students who used recall remembered 25 percent more of the text a week later.[6] Another study showed that recalling material one time doubled the long-term retention, while repeated recall resulted in a 400 percent improvement in retention when compared to studying one time.[7]

We've alluded to using recall during reading briefly before, but here's a more concrete rundown. Read a page as carefully as you can, trying to pick out the key ideas. Then look away and either tell yourself or write down those key ideas. If that's difficult, reread the text, (yes, it's okay to reread—but this time, try to pay closer attention!) then try again. Another attempt at recall the next day can be even more helpful; this allows you to check whether you've placed the information in long-term memory. (Again, if the material is particularly dense, you may want to reread it the next day prior to attempting recall.)

You can also do this with your notes. Rather than rereading your notes, try to *recall* what's in them. Or write summaries you can refer to later to study for the exam.

Think About the Text

To develop a deep understanding of what you're reading, it's paramount to find a way to actively think about the content and connect it to what you already know. There are many ways you can do this. **Pause to think occasionally, or to summarize what you read in your own words. Use the elaboration technique introduced in chapter 3, or find someone that you can discuss the content with.** Answer questions about the text. If your reading involves technical material, practice with problem sets. These ways of actively thinking about the material will not only improve understanding, but also make you more likely to remember details, since memory and understanding are connected.[8]

An Excellent Strategy for Active Reading: Annotation

Writing down short comments and questions as you read—"annotating"—is an excellent strategy for active reading that also makes it easier to review and refer back to the text.[9] If you read a document digitally, you can use the commentary/annotation tool provided. If you read on paper, you can make small notes directly in the margins, on Post-it Notes, or on a separate note sheet. This is what you should write down:

- Important ideas, rephrased using as few words as possible

- Relationships between concepts

- Your own examples or references

- Information you do not understand or need clarified

- Summaries of key paragraphs

- Potential test questions

The most important part of this strategy is to make sure you're putting the notes in your own words. By processing the information enough to paraphrase, you're gaining a deeper understanding of the material than if you simply rewrite what is in the text.

Make sure you include details as well as big-picture concepts. You need both to gain a deep understanding of the material. But don't just write a vague abstract summary, such as "identifies different types of leaves." Instead, be concrete: "5 types of leaf edges: entire = smooth; sinuate = curves/waves . . ."

When you have completed annotating the reading assignment, write **a three- to five-sentence summary** of the assignment. If you're unable to articulate the key ideas, you should review your annotations and attempt the summary again. If you're still unable to summarize the reading, you should reread the unclear sections and take additional notes on the reading.

How to Handle Excessive or Difficult Texts

Olav once coached "Nina," a college psychology student who felt overwhelmed by all her readings. Together, the pair piled up all her books and articles and discovered it would take three months of full-time reading to get through the material. No wonder Nina felt behind on her reading! She therefore categorized her readings into "Need to read" and "Nice to read." This allowed her to prioritize her studies—and ultimately, to be far more successful.

If you're studying at a university level and are overwhelmed with text, you can do the same. Ask your instructor, or students who took the course last year, for advice on which readings to consider mandatory, highly recommended, or supplementary. In college, for example, books often cover far more than your course does, so identifying the key chapters or sections within chapters can focus your studies while saving you a lot of time. If it's still too much, split the reading with other students and exchange summaries.

If you're doing your reading to try to acquire new skills for a job, the same rules apply. Try to prioritize and at least pick up the main points. Don't try to assimilate every little detail. Instead, just put in the time you can doing your best at improving your mastery.

Whatever you're reading, if you're struggling with difficult explanations, see if you can find simpler ones. Ask a friend or tutor to explain the material to you. Watch a YouTube video. Or take a break to allow the diffuse mode to kick in.

• • •

At the end of the learning road, there's usually a test. In the next chapter we'll see how to finish off strongly when all your learning is assessed and graded.

KEY TAKEAWAYS
FROM THE CHAPTER

- **Trying to read more quickly is not the way to read more effectively.** If you try to increase your reading speed beyond what feels natural, your comprehension will suffer.

- **Preview the text before reading it in detail.** Seeing the big picture allows you to make sense of and remember the details.

- **Avoid passive rereading.**

- **Practice active recall.** As you read, periodically look away and see if you can recall the key point of a page. This will help cement these key ideas in mind as well as build your understanding.

- **Find ways to think about the text—it improves comprehension.** You can take small time-outs while reading, answer questions about the text, summarize it in your own words, or discuss it with others.

10

How to Win Big on Tests

When Barb studied electrical engineering, she once did every possible practice problem she could to prepare for a tough circuits test. It would have been difficult to be more prepared than she was—in fact, before the test, some of the class's top students had come to her so she could explain a few of the more difficult concepts.

Yet Barb flunked the test, getting all ten questions wrong, while almost everyone else in the class did well. She thought, "They must be quicker on tests—or just plain smarter than me."

But she was wrong. It was just that there was *one* little test-taking trick she didn't know.

The Importance of Test-Taking Skills and Knowledge

We tend to think that to earn a high grade, all we need to do is learn the material well. But unfortunately, that's no guarantee. To do well on tests, we also need to be good at taking tests. That involves having knowledge about the upcoming test, and the ability to use smart test-taking strategies.

Barb learned the importance of this after she failed the circuits test. It turned out that the professor had wanted students to make

IF YOU HAVE A FACE-TO-FACE CLASS, ASK QUESTIONS

Those who ask their instructor questions get extra information, and those who don't, won't. But don't ask questions like "What will be on the test?" which many instructors find annoying. Instead, show evidence that you've already worked on whatever you're asking about. For example, you might ask, "I was reviewing my notes and the slides—I'm expecting these types of problems and prioritizing these topics for my studying . . . Does that seem to fit with what you would suggest?"

a special assumption* on the test. But he hadn't mentioned this assumption while teaching the class, it wasn't in the assigned chapters of the book, and it wasn't in the test instructions. However, without making this assumption, you couldn't get a single question right.

So why did the other students do so well and not Barb? It turned out that many of the other students had access to the prof's old tests. They knew roughly what the professor was going to be covering—and that the special assumption would probably be on the test. After discovering what the other students were up to, Barb started interacting more with her fellow students to make sure she had access to the same test information.

So before you start preparing for a test, do your best to find out what it will look like. Not the exact questions (we're not advocating that you cheat!), but what type of questions you'll get, how they'll be evaluated and graded, and any expectations and assumptions. There are several ways to familiarize yourself with an upcoming test, such as reviewing information you have about the test, clarifying things you're uncertain about, and discussing the upcoming test with others. But perhaps the most important is to practice with old tests from that instructor, if at all possible.

*A 0.7 volt drop across the diodes.

The Importance of Practicing with Old Tests

Test-taking is invaluable for learning. One hour of test-taking, for example, will teach you far more than one hour of studying.[1] And taking practice tests are valuable too—in fact, **research shows that the *best* way to prepare for a test is by practicing with problems and questions that are similar to those of the upcoming test.**[2] It's easiest to do this by working with old test questions. Of course, problems from other sources can be nearly as helpful—although they may not include the exact type of questions you're likely to get on the test.

But remember: practice problems won't help if you just look at the answers. This just loads the information into working memory, where it can easily fall out. **You need to work practice problems yourself, even if you think you already understand them. This working out *yourself* of practice problems is what takes you to the "pro learning" level.**

Where do you find old tests? When it comes to licensing and professional exams, there are often old versions of the test available, either in books or online. For college and online classes, a search for the name of your class/unit, a key term you know will be covered, and practice/quiz/test/example/questions can also unearth surprisingly helpful materials. You could also refer to comprehensive websites like Course Hero for practice questions. Just be careful that you're not breaking any rules by accessing and using prior materials.

If you can't find old practice questions, come up with some questions yourself that you think are likely to appear on the test. If you're taking a course with stated learning objectives, try rewriting them as exam questions. This can work especially well for psychology, history, and other largely nonquantitative courses.

You may have realized that taking practice tests is important for another reason—as with other types of practice, it helps develop the long-term memory links of your procedural system. Remember, having links in both your declarative and procedural systems gives you a much broader and richer understanding of the materials. You will also be able to react more quickly and intuitively when you're facing the real test.

Plan Your Time

When Olav received the exam schedule for his first semester as a graduate student at Oxford, he felt certain that he would fail all his exams. At his undergraduate school in Norway, lectures would normally end one to two months before the final exams, leaving students with plenty of time to prepare. As Olav looked down at his first exam schedule in Oxford, he realized he had only one week to study for five exams. What should he do?

Olav took comfort in the fact that he wasn't the only one with little time to prepare. (And if it *was* too little time to prepare, he was pretty sure the university wouldn't fail an entire batch of nearly three hundred students.) Next, he set up a *plan*, so that he could get the most out of whatever time he had left.

A plan doesn't create more time, but it does allow you to spend your time wisely. A plan will also reduce stress, since you don't have to worry about what to do. All you have to do is follow the plan.

OVERALL STUDY PLAN

W	Mon	Tue	Wed	Thu	Fri	Sat	Sun
21	History	History	Math	Math	Spanish	Day off	Day off
	Math	Spanish	Spanish	History	History		
22	Spanish	History	History	Math	Math	Math	Day off
	Math	Math	Spanish	Spanish	Spanish	Spanish	
23	Math	Math	Math Exam	Spanish	Spanish Exam	History	History
	Math	Math		Spanish		History	History
24	History Exam						

If you have multiple tests coming up, start by finding out how much time you have to prepare for each test. Look at the number of days available for preparation and divide them among the exams. Next, use a calendar to plan which days you'll prepare for the different exams. Try to prepare for at least two exams on any given day to build spacing into your study. The only exception is the one to two days right before a test. That's when focusing only on the immediate subject, if you're able to, can be beneficial.

When you know how much time you have to prepare, and when you'll do it, it's smart to think carefully about how you'll spend that time. What parts of the curriculum will you focus on? How much time will you spend solving past exam problems, reading, reviewing notes, and discussing with peers? Sum your thoughts up in a quick study plan. If you want, you can go into more detail for a specific subject.

DETAILED STUDY PLAN FOR SPANISH

Tue	Solve past exam questions (4hr) Study group (2hr)
Wed	Study chapters 1-2 (3hr) Make notes (2hr)
Fri	Study chapters 3-4 (3hr) Make notes (2hr)
Mon	Study chapters 5-6 (3hr) Make notes (2hr)
Wed	Study group (2hr) Past exam questions (3hr)
Thu	Past exam questions (4hr) Review notes (1hr)
Fri	Past exam questions (4hr) Review notes (1hr)
Sat	Past exam questions (4hr) Review notes (1hr)
Thu	Past exam questions (7hr) Review notes (3hr)
Fri	Exam

All this is similar to what you would do, say, if you were studying for an MBA while working a full-time job and juggling family responsibilities. In this case, buy-in from your family matters as well.

You'll be making sacrifices and will probably suffer from undesirable lack of sleep on occasion, but that's just the way of things. Do the best you can, realizing that there is no optimal strategy, but that planning your time as carefully as you can is important.

It's common to be overly optimistic, so it's okay to change your plan as you go along and get a better feel for how long your studies will take. Incidentally, the value of planning isn't the completed plan as much as it is the reflections and considerations you have to do in order to create the plan.

Do your best to set aside enough time for sleep, especially in the days leading up to the test. Even if you struggle to fall asleep on the night before a test, you can tackle test day much better if you've stored several nights of good sleep already.

Read the Instructions, and Each Question, Carefully

Olav once had to take a computerized multiple-choice test to qualify for a job interview. The test had thirty questions about a text and its related numbers and graphs. Olav skimmed the instructions and hit the start button. It all went well at first, but as he neared the end of the allotted time, many questions remained. He didn't want to miss potential points, so he randomly selected the remaining answers. Once the test was submitted, Olav reread the instructions. That's when he discovered that the test had been designed so test-takers couldn't answer all questions. The instructions warned against guessing—wrong answers meant a lot of negative points.

In other words, Olav had just taken one of those rare tests in which a wrong answer was much worse than a blank answer. Of course, he failed the test—not because of his qualifications, but because he hadn't read the instructions carefully.

People misread and misunderstand all the time. On tests, however, this can be devastating. **Be sure to read any test instructions carefully.** If a test has five essay questions, you need to know whether you're to answer all five or only one. It can also be wise to look at each question three times: before you start your answer, midway through, and when you're done with your answer. This will help you stay on track as you're answering. Sometimes, our understanding of a question improves or changes once we have activated our knowledge, which typically happens when you have worked at a problem for a bit.

Keep Track of Time

Olav used to struggle with completing tests. He would typically run out of time toward the end of a test and sometimes leave significant parts unanswered. On one such occasion, he left the last question on a five-question math exam unanswered. It didn't matter that he had answered four questions well. It couldn't make up for the fact that he left 20 percent of the exam unanswered, which significantly reduced his grade on the test.

To avoid running short on time, it can be wise to have a mental time plan to follow, and to check the clock several times throughout the test to know if you're on track. One approach is to divide the total test time by the number of questions on the test, adjusting for any questions that are weighted more. That will give you a rough benchmark of how much time you have for each question.

IT'S ALL IN THE TIMING
Taking practice tests—and timing them to be similar to the amount of time you will have on the real test—is one of the best ways to make sure you're on track time-wise during the real test.
  Practice tests can be a dry run to help give you a better idea of what might take longer or be more difficult.

Say you have 60 minutes to answer ten questions. That's six minutes per question. When you're halfway through the time, that means you should have completed five or six questions. If you've only answered two questions in 30 minutes, you're heading for trouble.

If you aren't disciplined enough to make yourself stop working on a question and carry on to the next, the risk is that you will spend way too much time on some questions at the expense of others. Although you might have written a fantastic answer to some questions, it's unlikely to make up for incomplete or blank answers on other questions (as Olav painfully experienced).

Keep in mind the Hard Start approach: start with what seems to be the hardest problem—but only work on it for a few minutes, until you get stuck. When you move to an easier problem, your diffuse mode can work in the background on the harder problem.

The discipline you use to pull yourself from a problem in the Hard Start Technique is the same discipline you will use to keep yourself on schedule with answering questions. You don't want any one question to hook you in too completely!

Review Your Answers

If you have some time at the end of the test, use it to review your answers. Check to see whether your diffuse mode has anything to add or whether it nudges you to change any answers. Interestingly, research has found that when test-takers do change their answers during reviews, the changes are generally from incorrect to correct.[3] This is also a good time to catch and correct ambiguous language. Another tip is to note down your most common types of errors so you can check your work with more specific focus.

As you do the final review, make sure that you don't leave hard questions blank, unless a wrong or incomplete answer gives you negative points. Instead, write down whatever relevant information you can think of. For example, you might note down the steps you would have taken or the formula you should have used. This can give you partial credit. Students sometimes neglect to provide answers if all they have are definitions or what they assume to be common knowledge. That may well be exactly what is being sought.

Dealing with Test Anxiety

Being a bit anxious or stressed before an important test is normal and actually *beneficial* for a good test performance.[4] So instead of feeling that you're stressed because of the test, reframe it as "I'm stressed because I'm gearing up to do my best!"

IF POSSIBLE, CHOOSE COURSES STRATEGICALLY

When Barb studied engineering, she used to select her classes based on final exam dates to avoid having, for example, two consecutive final exams in a day. She found that being able to space out final exams over as many days as possible gave her more time to prepare and made exams less stressful. She also preferred courses that had exams lasting three hours instead of, say, one hour. Why? Barb noticed that the longer exams usually allowed more time per question, taking away some pressure and allowing for more diffuse mode thinking. **Being a bit strategic when you plan your courses, if you can, can help you earn better grades.** For example, if you typically perform better in classes where the grade is based on many small assignments, as opposed to one final exam, then that's something to consider when you choose classes.

Some "night owl" students struggle in the mornings—it's wiser for them to aim for late afternoon or evening sections. Others do best with online video lectures that they can stop and restart if they lose track of an explanation. It helps to analyze what specifically worked for each course that turned out well, and what didn't work for courses that didn't turn out so well.

You'll put more effort into well-designed courses with good instructors. This means that getting "intel" on courses from other students or websites such as Class Central and Rate My Professors can be beneficial. Keep in mind, however, that sometimes students can give unworthy ratings when they don't get the grade they want.

In the end, remember that although grades matter, what you have learned and what you will learn will matter far more. The real prize of an education is to become an effective lifelong learner.

If you suffer from serious anxiety, creating strong sets of links in both declarative and procedural memory by using the study techniques we've shown you in this book is the best way to reduce that anxiety and do well on tests.[5] In the stressful time leading up to an exam, it's also better to focus on *process* goals (such as completing three hours of studying) over end goals (such as getting an A). Focusing on the *process* will take away some pressure and make you more likely to reach your end goal anyway.

Just as the test is being handed out or you hit "start" for an online exam, you may feel your anxiety turning to panic. This can happen because while stressed, you might tend to breathe shallowly, only drawing air into the top of your chest. You don't get enough oxygen, and your body panics. To tackle this, watch your breathing. Just before you might go into panic mode, put your hand on your belly and try to draw air so deeply into your lungs that your hand moves up and down. This deep breathing can allow you to grow calmer and steadier.

• • •

In the next and final chapter, we'll look at how you can get the most out of the tools in this book and how to continue to improve your learning.

KEY TAKEAWAYS
FROM THE CHAPTER

Before the test:

- **Familiarize yourself with the test and the type of questions it will likely ask.**
  - Review information about the test.
  - Clarify things you're uncertain about.
  - Discuss the upcoming test with others.

- **Practice with as many previous test questions from old tests as possible.** Don't just look at the answers—that simply loads the information into working memory, where it can easily fall out.

- **Make a test preparation plan** that specifies how much time you will spend and what you're going to focus on to prepare for the test.

- **Choose courses strategically.** Remember that the real prize of education is to become an effective lifelong learner.

During the test:

- **Read test instructions and questions carefully.** Rereading the question when you're already a little into your answer can improve your interpretation of it.

- **Keep track of time on tests.** Know the approximate amount of time you have for each section of the test and check the time regularly to ensure that you're on track.

- **Spend whatever time you have left reviewing your answers.** Make sure you have answered all questions (unless guessing is penalized) and that your answers are clear and include all the important points.

- **Remember the Hard Start Technique.** Where possible, start with the most difficult questions or problems first. If stuck, stop, continue with the rest of the exam, and come back to the difficult question later. Let that diffuse mode work for you!

- **Moderate amounts of stress before and during a test can actually be advantageous.** Be sure to breathe slowly and deeply, down into the lower parts of the chest, just before the test to ensure you're calming your autonomic nervous system.

How to Be a Pro Learner

"You can't sing," the judges repeat. The contestant gapes, unable to believe what he's hearing, then angrily drops the microphone and storms offstage.

If you've ever watched a talent show such as *The X Factor* or *America's Got Talent,* you've probably seen this happen. But here's the important question: Why do so many people think they're so good at something when they obviously are not?

Let's pause for a moment to look at what we've covered. So far, we've shared some of the best mental tools that science and experience have taught us, such as:

- Using the Pomodoro Technique and setting up a distraction-free environment to beat procrastination.

- Taking breaks and using the Hard Start Technique on tests to overcome being stuck.

- Studying actively: testing yourself, elaborating, interleaving, and spacing out your sessions in order to form strong sets of neural connections in both your declarative and procedural systems.

- Using acronyms, sentences, vivid images, the Memory Palace, and internalization to memorize and internalize.

- Planning when, where, and how to respond to obstacles in order to exert more self-discipline.

- Finding value, ensuring mastery, and setting goals to motivate yourself.

- Previewing, practicing active recall, and annotating in order to read effectively.

- Analyzing and practicing with previous tests and monitoring your time to earn a good grade on exams.

These are all powerful tools that, if applied correctly, will help you become a better learner. But how do you ensure that you apply the right strategies at the right time and in the right manner? And how do you adapt these strategies to new situations outside your studying? How do you *really* learn like a pro?

For this, you need what researchers call metacognition.

The Importance of Metacognition

You can think of metacognition as an extra brain outside your main one. This extra brain thinks about *how* you are thinking (meta-cognition means "thinking about thinking"). This extra brain pauses to consider how you should best approach problems and what strategies you should use. It stops you midway through a session (perhaps several times) to assess whether the approach you are using is working. And at the end of a session, your

extra brain looks back to see if you should do something differently. In other words, metacognition is what allows you to learn *when* to use the tools, and improves your ability to *use* those tools. It's critically important in becoming a pro learner.

And that extra brain is also critical to avoid being surprised by harsh judgments in talent shows. Why are some contestants so surprised? In addition to lack of talent, they most likely lack proper metacognition. Had they been metacognitive, they would have actively sought out objective feedback and tried to assess themselves from a distance, accepting and learning from criticism.

Why are some people more metacognitive than others? Research indicates that, no surprise, people who are overconfident in their own abilities have poorer metacognition.[1] But anyone can become more metacognitive with practice.

Asking Yourself Metacognitive Questions

The easiest way to become more metacognitive is simply to start asking yourself higher-level questions, such as:

- What are the resources available to help me when I struggle?

- Do I focus on the right things at the right level in my studying? Should I prioritize differently?

- Can I be a more effective student? What can I improve?

- What do I find difficult and why?

A Model for Being a Metacognitive Learner

A more structured way to become a metacognitive learner is to follow this four-step model for self-regulated learning, based on the work of Canadian psychologists Phil Winne and Allyson Hadwin.[2]

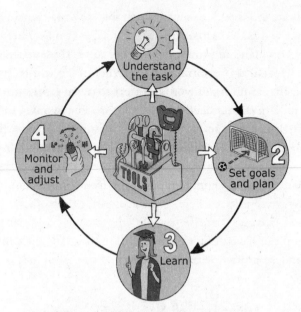

The four steps of being a self-regulated learner.

Step 1: Understand the Task

Understand what's required of you, how you'll be assessed, and the time and resources you have available. If you're studying a topic at a company workshop, online, or in college, you may use learning objectives related to that topic to guide you to exactly what you need to learn. If you're preparing for an exam, you might analyze past exams to better understand what kinds of questions you will likely have to answer. You could also ask your instructor and peers for their understanding of the task. If you don't know where to start on a task, that's typically a sign that you haven't understood the task well.

Step 2: Set Goals and Plan

Think about your ambition—at what level you want to perform— and then break your task down into concrete goals. Once you have set some goals, plan when, where, and how you will work to achieve them. That includes selecting appropriate learning techniques. If you're studying a list of Italian vocabulary, you may set yourself a goal of memorizing 100 percent of the words, and plan to make flash cards for the words and use retrieval practice 10 minutes a day for five days, spaced out over a period of two weeks. If you're studying photosynthesis, you may set yourself a goal to be able to understand the process so well you can explain it to anyone. To achieve this, you may plan to read about photosynthesis from three different sources and practice using the elaboration technique.

Step 3: Learn

Learn according to your plan, experimenting with different tools.

Step 4: Monitor and Adjust

As you're learning, make sure to take a step back regularly to assess how your studying is going. Assess whether you're actually progressing in your learning, and whether you're using the best learning strategies. You might need to adjust the way you're learning, and that's perfectly fine. In fact, changing practices when progress is slow or nonexistent is an important part of being a self-regulated learner. For example, if you realize that you're struggling to understand a difficult text, you might want to try studying another source, such as an explainer video on YouTube or an online course.

This four-step process is recursive, meaning that you'll likely go through the steps several times on your way to mastering a difficult subject. Once you've been through a cycle, you often have a better understanding of the task, and you may have learned some things that make you update your goals and plan. Based on that, you change the way you study, and as you progress, you may improve your task understanding and make further changes to goals and plans.

Practicing to become more self-regulated in your studying can really pay off: one study found that training students to become self-regulated learners could move them from the fiftieth to the seventy-fifth percentile.[3]

Learning from Past Exams

One way to evaluate your performance (step 4) on tests is to use a structured form like the one shown below. Such a form forces you to think about the different aspects of the test-taking experience and to reflect on what you could have done better.

WHAT YOU DID BEFORE THE TEST

Question	Yes	No	What can you do differently next time to improve?
Did you acquire enough information about the exam beforehand to know what and how to study?			
Did you set aside enough time to be able to prepare properly?			
Did you study all the relevant parts of the curriculum?			
Did you work effectively and in a focused fashion during the time you had set aside for studying?			

WHAT YOU DID DURING THE TEST

Question	Yes	No	What can you do differently next time to improve?
Did you understand the instructions and questions well?			
Did you answer all the questions?			
Were you tired or hungry so that you could not concentrate well?			

Question	Yes	No	What can you do differently next time to improve?
Did you panic or experience severe anxiety so that you could not concentrate well?			
Did you distribute the time well among the different questions?			
Did you make any careless mistakes?			
Did you remember the main ideas for each question?			
Did you remember the details for each question?			
Did you organize your response in a clear and cohesive way?			

A Final Word

Your brain is the most valuable and complex tool you possess. We hope that this book can help you get more out of it, and that it has inspired you to improve the way you learn. So let's end this chapter, and book, with a metacognitive question:

Write it down: What are your biggest takeaways from this book? What will you do differently in your studies going forward?

KEY TAKEAWAYS
FROM THE CHAPTER

- To become an effective learner, you need to apply the right learning tools at the right time and constantly think about how you can improve.

- To do this, you need metacognition—a brain outside your brain—that takes a step back to pause and ask high-level questions.

- Ask yourself metacognitive questions, such as:
  - What are the resources available to help me when I struggle?
  - Do I focus on the right things in my studying? Should I prioritize differently?

- Use the four-step self-regulated learning model to make a habit of being metacognitive:

  Step 1: Understand the task.

  Step 2: Set goals and plan.

  Step 3: Learn.

  Step 4: Monitor and adjust.

Checklist: How to Become an Effective Learner

Below, we have summarized the advice in this book. (Yes, you need to have read the book to fully benefit from it).

1. **Focus intently and beat procrastination.**

   ☐ Use the Pomodoro Technique (remove distractions, focus for 25 minutes, take a break).

   ☐ Avoid multitasking unless you find yourself needing occasional fresh perspectives.

   ☐ Create a ready-to-resume plan when an unavoidable interruption comes up.

   ☐ Set up a distraction-free environment.

   ☐ Take frequent short breaks.

2. **Overcome being stuck.**

   ☐ When stuck, switch your focus away from the problem at hand, or take a break to surface the diffuse mode.

   ☐ After some time completely away from the problem, return to where you got stuck.

   ☐ Use the Hard Start Technique for homework or tests.

   ☐ When starting a report or essay, do not constantly stop to edit what is flowing out. Separate time spent writing from time spent editing.

3. Learn deeply.

☐ Study actively: practice active recall ("retrieval practice") and elaborating.

☐ Interleave and space out your learning to help build your intuition and speed.

☐ Don't just focus on the easy stuff; challenge yourself.

☐ Get enough sleep and stay physically active.

4. Maximize working memory.

☐ Break learning material into small chunks and swap fancy terms for easier ones.

☐ Use "to-do" lists to clear your working memory.

☐ Take good notes and review them the same day you took them.

5. Memorize more efficiently.

☐ Use memory tricks to speed up memorization: acronyms, images, and the Memory Palace.

☐ Use metaphors to quickly grasp new concepts.

6. Gain intuition and think quickly.

☐ Internalize (don't just unthinkingly memorize) procedures for solving key scientific or mathematical problems.

☐ Make up appropriate gestures to help you remember and understand new language vocabulary.

7. Exert self-discipline even when you don't have any.

☐ Find ways to overcome challenges without having to rely on self-discipline.

☐ Remove temptations, distractions, and obstacles from your surroundings.

☐ Improve your habits.

☐ Plan your goals and identify obstacles and the ideal way to respond to them ahead of time.

8. Motivate yourself.

☐ Remind yourself of all the benefits of completing tasks.

☐ Reward yourself for completing difficult tasks.

☐ Make sure that a task's level of difficulty matches your skill set.

☐ Set goals—long-term goals, milestone goals, and process goals.

9. Read effectively.

☐ Preview the text before reading it in detail.

☐ Read actively: think about the text, practice active recall, and annotate.

10. Win big on tests.

☐ Learn as much as possible about the test itself and make a preparation plan.

☐ Practice with previous test questions—from old tests, if possible.

☐ During tests: read instructions carefully, keep track of time, and review answers.

☐ Use the Hard Start Technique.

11. Be a pro learner.

☐ Be a metacognitive learner: understand the task, set goals and plan, learn, and monitor and adjust.

☐ Learn from the past: evaluate what went well and where you can improve.

Acknowledgments

Many people have helped us in the process of writing this book, and we are grateful to each and every one. First and foremost, we'd like to thank the more than one hundred beta readers who read through the manuscript and provided detailed feedback. It made the book much stronger. We'd like to thank our fantastic editor, Daniela Rapp, the Empress of Editors. We would also thank our talented illustrator, Oliver Young, for finding creative ways to illustrate key concepts.

Notes

1. How to Focus Intently and Beat Procrastination

1   **Disruption and multitasking are harmful:** Mokhtari et al., 2015.
2   This is a process called *consolidation*. Fiebig and Lansner, 2014. The fact that brief mental breaks help with consolidation is discussed in Wamsley et al., 2010.
3   **Pain in the brain:** Lyons and Beilock, 2012.
4   **52 minutes with a 17-minute break:** Thompson, 2014.
5   **Cognitive cost of using a cell phone during breaks:** Kang and Kurtzberg, 2019.
6   **Students not using phones did better:** Kuznekoff and Titsworth, 2013.
7   **Phones are distracting when nearby:** Ward et al., 2017.
8   **Best to leave phone out of reach:** Cutino and Nees, 2016.
9   **Neural information made available to consciousness:** Dehaene and Changeux, 2011.
10   **The cognitive cost of task-switching:** Rubinstein et al., 2001. **Supertaskers:** Medeiros-Ward et al., 2015.
11   **Task-switching reduces cognitive fixation and increases creativity:** Kapadia and Melwani, 2020; Lu et al., 2017.
12   **An average of 35 seconds:** Mark et al., 2016.
13   **Websites blocked:** Mark et al., 2017.
14   **Ready-to-resume plan:** Leroy and Glomb, 2018.
15   **Effects of long periods of focus:** See Garrison et al., 2015, for an interesting study of the effects of focused meditation on reduction in activity of the default mode network, which plays an important role in creativity.
16   **Cognitive exhaustion plays a role:** Madjar and Shalley, 2008. "Cognitive exhaustion" is an elusive term—there don't appear to be any fMRI studies that give a metabolic sense of what might actually be happening when you tire mentally. Theories of failures of self-control as arising from "ego depletion" have been questioned. Carter et al., 2015. At the same time, it's pretty clear that there is a metabolic cost to the areas of the brain that are used more during thinking processes. Ampel et al., 2018.
17   **Do nothing when taking a break:** Wamsley, 2019.
18   **Hours spent studying by medical school students:** Liles et al., 2018.
19   **Average hours spent studying by more typical students:** Bart, 2011.
20   **Breaks allow new ideas to settle into place:** Wamsley, 2019.
21   **Working memory and music:** Christopher and Shelton, 2017.
22   **Music benefits those with ADHD:** Antonietti, et al, 2018.
23   **History and overview of binaural beats:** Turow and Lane, 2011.
24   **A meta-analysis of binaural beats:** Garcia-Argibay et al., 2019.
25   **Meta-analyses of the effects of meditation:** Chiesa et al., 2011; Sedlmeier et al., 2012.
26   **Yoga's effects on the brain:** Gothe et al., 2019.

2. How to Overcome Being Stuck

1 *Focused* and *diffuse* modes are referred to by psychologists as task-positive and task-negative networks, respectively. Neuroscientists use the term *default mode network* for the diffuse mode. Fox et al., 2005.
2 **Laying pathways:** Sekeres et al., 2017.
3 **The default mode ("diffuse mode") and creativity:** Kühn et al., 2014.
4 **Diffuse mode working behind the scenes:** Sio and Ormerod, 2009, but a number of factors may be at play, including the avoidance of cognitive fixation (next note).
5 **Alleviating a tendency to cognitively fixate on ineffective ideas or problem-solving strategies:** Lu et al., 2017.
6 **Coffee shop:** O'Connor, 2013. **Noise disrupts focus (working memory):** Sinanaj et al., 2015.

3. How to Learn Anything Deeply

1 You do something more than just strengthen connections when you practice over many days—you also allow for a process called myelination, which coats your neurons with an insulation that helps speed up neural signals. But the concept of myelination is not necessary to understand the creation of new links that initiates the learning process. Excellent review papers on the creation of links in memory are Poo et al., 2016, and Josselyn and Frankland, 2018.
2 **Active practice reduces test stress:** Smith et al., 2016.
3 **Retrieval:** Karpicke, 2012; Antony et al., 2017.
4 Renno-Costa et al., 2019, notes that "synapse strength is proportional to synapse size . . ." but of course there is also an increase in receptor density and many other factors that increase synaptic strength—the illustration is just meant to give a feel for the improved connectivity, rather than to be completely accurate from an anatomical perspective.
5 **Guessing answers:** Kornell et al., 2009
6 This is related to Anders Ericsson's conceptions of "deliberate practice." See Ericsson and Pool, 2016, for a readable, book-length description of Ericsson's research.
7 **Elaboration:** Dunlosky et al., 2013.
8 **Explaining steps:** Berry, 1983.
9 Understanding not only concepts but also *differences* between concepts appears to relate to hippocampal differentiation as well as integration of material. Sekeres et al., 2018.
10 **Interleaving and the study of artistic styles:** Kornell and Bjork, 2008.
11 **Coaching and interleaving:** Beilock, 2010.
12 **BDNF and exercise:** Szuhany et al., 2015. Image after Lu et al., 2013.
13 **No current guidelines:** Per email on August 16, 2019, with Jennifer Heisz, associate director, Physical Activity Centre of Excellence, McMaster University.
14 **10 percent improvement:** Heisz et al., 2017.
15 **Meta-analysis:** Chang et al., 2012.
16 **Exercise guidelines in the United States:** U.S. Department of Health and Human Services, 2018.

17 **Exercise and physiological effects:** Basso and Suzuki, 2017; Szuhany et al., 2015.
18 **High-intensity interval training:** Jenkins et al., 2019.
19 **Music helpful:** Stork et al., 2019.
20 **Haven't panned out: Ginkgo:** Laws et al., 2012. **Ginseng:** Geng et al., 2010.
21 **Caffeine:** Glade, 2010; Nehlig, 2010. **Phytochemicals and guarana:** Haskell et al., 2013.
22 **Glucose:** Smith et al., 2011.
23 **Overeating versus fasting in relation to cognition:** Mattson, 2019.
24 **Fasting:** Brandhorst et al., 2015; Mattson, 2019.
25 **Cocoa:** Socci et al., 2017. **Flavonoids:** Rendeiro et al., 2012. **Curcumin:** Cox et al., 2015.
26 **Enhancement of combined effect:** Adan and Serra-Grabulosa, 2010.
27 **Exercise and diet combine to enhance cognition even more when done in conjunction:** van Praag, 2009.
28 **Vegetables, nuts, and berries and cognition:** Miller et al., 2017; Pribis and Shukitt-Hale, 2014.
29 **Avoid fast foods:** Tobin, 2013.
30 **Reopening of critical period of learning and promotion of neural plasticity:** Gervain et al., 2013; Ly et al., 2018.
31 **Synthetic stimulants:** Repantis et al., 2010; Smith and Farah, 2011.
32 **Neuroscientists rarely use noninvasive brain stimulation on themselves:** Shirota et al., 2014.
33 **Don't remain committed users:** Jwa, 2018.
34 **Sleep and budding spines:** Yang et al., 2014.
35 **Toxins get cleared out:** Xie et al., 2013.
36 **Naps help:** Cousins et al., 2019.
37 **Lots more about sleep can be found in** Walker, 2017.
38 **Advice about blue light:** Harvard Medical School, 2012, updated 2018.
39 **Put cell phone in another room:** Hughes and Burke, 2018.
40 **The efficacy of weighted blankets:** Ekholm, B, et al. "A randomized controlled study of weighted chain blankets for insomnia in psychiatric disorders." *Journal of Clinical Sleep Medicine* 16, no. 9 (2020): 1567–1577.
41 Information on relaxing to sleep based on Winter, 1981.
42 **Various aspects of breath and breathing:** Nestor, 2020.

4. How to Maximize Working Memory—and Take Better Notes

1 **Definitions of working memory can differ dramatically:** Cowan, 2017.
2 **Three or four pieces of information in working memory:** Cowan, 2001; also see much more up-to-date work in this area by Cowan and others.
3 **Voice and pictures presented simultaneously make it easier to learn:** Mayer, 2014.
4 **General sense from research:** Jansen et al., 2017; Zureick et al., 2018.
5 This is actually a modified version of Cornell note-taking.
6 **Medical students and same-day review:** Liles et al., 2018.

7   **Retrieval practice more effective than concept mapping:** Karpicke and Blunt, 2011.

8   **Rewatching video lectures instead of taking notes isn't a good idea:** Liles et al., 2018.

9   **Dr. David Handel:** Handel, 2019.

10   **Students who use notes from others do almost as well:** Kiewra et al., 1991.

5. How to Memorize

1   **Freeing up working memory helps you think more easily about complex topics:** Sweller et al., 2011.

2   French and Russian revolution example from Agarwal and Bain, 2019.

3   **Conceptual understanding, procedural fluency:** Karpicke, 2012; Rittle-Johnson et al., 2015.

4   **Individual pieces of learned data help the learner to generate a big picture understanding;** Schapiro, et al., 2017. **Big picture conceptual understanding (schemas) help a learner to learn new individual pieces of information:** Ghosh and Gilboa, 2014.

5   **Your brain is super visual:** Oakley and Sejnowski, 2019.

6   **2,560 pictures:** Standing et al., 1970.

7   **Vivid images are easier to remember:** D'Angiulli et al., 2013.

8   **Neural reuse theory:** Anderson, 2010.

6. How to Gain Intuition and Think Fast

1   **Two completely different systems:** See chapter 6, Oakley et al., 2021, for a complete discussion and set of research citations related to the declarative/procedural model of learning.

2   "When students are convinced that they know the answer to specific questions, 54–64% deliberately choose not to re-engage with the questions . . . and likely never revisit those questions even if they have opportunities to do so": Pan and Bjork, in press.

3   **Actively pulling the information from your own memory aids the development of intuition:** Himmer et al., 2019.

4   The value of interleaving was first discovered in relation to the procedural learning system. See Pan and Bjork, in press.

5   **Internalization includes retrieval processes:** Adesope et al., 2017; Agarwal et al., 2008.

6   **Broader problem-solving spurs analogical transfer:** Pan and Bjork, in press.

7   **Declarative/procedural model and language learning:** Ullman and Lovelett, 2016.

8   **Delay of review for several months:** Cepeda et al., 2008.

9   **Sleep and mind wandering help during spaced repetition:** van Kesteren and Meeter, 2020.

10   **Making gestures helps students learn foreign language words better:** Macedonia et al., 2019; Straube et al., 2009.

7. How to Exert Self-Discipline Even When You Don't Have Any

1   **Benefits of self-discipline:** Duckworth et al., 2019; Hofmann et al., 2014; Moffitt et al., 2011.
2   **Removing temptations:** Milyavskaya and Inzlicht, 2017.
3   **The time it takes to form a habit:** Lally et al., 2010.
4   **Students who planned studied 50 percent longer:** Sheeran et al., 2005.
5   **Planned responses to obstacles completed 60 percent more practice questions:** Duckworth et al., 2011.
6   **91 percent of people who planned met their exercise goals:** Sniehotta et al., 2006.
7   **Roosevelt best-read man in America:** Roosevelt, 1915.

8. How to Motivate Yourself

1   **Rats, dopamine, and increased effort:** Bardgett et al., 2009.
2   **Humans will work harder when dopamine levels are high:** Treadway et al., 2012; Wardle et al., 2011.
3   **Thinking about benefits motivates:** Hulleman et al., 2010.
4   **No mental break when you reach for a mobile phone:** Kang and Kurtzberg, 2019.
5   **Mental contrasting as motivational technique:** Oettingen and Reininger, 2016.
6   **SMART goals:** Doran, 1981.
7   **Humans have a need for relatedness:** Ryan and Deci, 2000.
8   **Motivation contagion:** Dik and Aarts, 2007.

9. How to Read Effectively

1   **The reading process, from letters to idea:** Leinenger, 2014; Hruby and Goswami, 2011; Rayner et al., 2016.
2   **Your mind processes information during jumps:** Irwin, 1998.
3   **Reading experts on Anne Jones reading a Harry Potter book in 47 minutes:** Rayner et al., 2016.
4   **Carol Davis quote:** email from Carol Davis, July 20, 2019.
5   **Recall:** Karpicke, 2012. We use the word "recall" as a synonym for the technical term "retrieval."
6   **Study that shows the effect of testing on reading retention:** Roediger III and Karpicke, 2006.
7   **Doubled retention:** Karpicke, 2012.
8   **Memory and understanding are connected:** Karpicke, 2012.
9   Thanks to Kristey Drobney for the ideas behind this section.

10. How to Win Big on Tests

1   **Positive effects of test-taking:** Rowland, 2014.
2   **Best way to prepare for tests:** Adesope et al., 2017.
3   **Answer changes during review:** Bridgeman, 2012.

4   **Test anxiety helps performance:** Brady et al., 2018.
5   **Reducing test anxiety:** Beilock, 2010; Karpicke and Blunt, 2011; Karpicke, 2012.

11. How to Be a Pro Learner

1   **Overconfident people have poorer metacognition:** Molenberghs et al., 2016.
2   **Self-regulated learning in four steps:** This model is a slight modification of the self-regulated learning model put forth in Winne and Hadwin, 1998.
3   **Effect of training students to become self-regulated:** Dignath and Büttner, 2008.

Bibliography

Adan, A., and J. M. Serra-Grabulosa. "Effects of caffeine and glucose, alone and combined, on cognitive performance." *Human Psychopharmacology* 25, no. 4 (2010): 310–17.

Adesope, Olusola O., et al. "Rethinking the use of tests: A meta-analysis of practice testing." *Review of Educational Research* 87, no. 3 (2017): 659–701.

Agarwal, P. K., and P. Bain. *Powerful Teaching: Unleash the Science of Learning.* San Francisco, CA: Jossey-Bass, 2019.

Agarwal, Pooja K., et al. "Examining the testing effect with open- and closed-book tests." *Applied Cognitive Psychology* 22, no. 7 (2008): 861–76.

Ampel, Benjamin C., et al. "Mental work requires physical energy: Self-control is neither exception nor exceptional." *Frontiers in Psychology* 9 (2018): Art. No. 1005.

Anderson, Michael L. "Neural reuse: A fundamental organizational principle of the brain." *Behavioral and Brain Sciences* 33, no. 4 (2010): 245–66.

Antonietti, A., et al. "Enhancing self-regulatory skills in ADHD through music." In *Music Interventions for Neurodevelopmental Disorders,* edited by Alessandro Antonietti, et al., 19–49 New York, NY: Springer, 2018.

Antony, J. W., et al. "Retrieval as a fast route to memory consolidation." *Trends in Cognitive Science* 21, no. 8 (2017): 573–76.

Bardgett, Mark E., et al. "Dopamine modulates effort-based decision making in rats." *Behavioral Neuroscience* 123, no. 2 (2009): 242–251.

Bart, Mary. "Students study about 15 hours a week, NSSE finds." *The Faculty Focus* (2011). https://www.facultyfocus.com/articles/edtech-news-and-trends/students-study-about-15-hours-a-week-nsse-finds/.

Basso, Julia C., and Wendy A. Suzuki. "The effects of acute exercise on mood, cognition, neurophysiology, and neurochemical pathways: A review." *Brain Plasticity* 2, no. 2 (2017): 127–52.

Beilock, Sian. *Choke: What the Secrets of the Brain Reveal About Getting It Right When You Have To.* New York, NY: Free Press, 2010.

Berry, Dianne C. "Metacognitive experience and transfer of logical reasoning." *Quarterly Journal of Experimental Psychology Section A* 35, no. 1 (1983): 39–49.

Brady, Shannon T., et al. "Reappraising test anxiety increases academic performance of first-year college students." *Journal of Educational Psychology* 110, no. 3 (2018): 395–406.

Brandhorst, Sebastian, et al. "A periodic diet that mimics fasting promotes multi-system regeneration, enhanced cognitive performance, and healthspan." *Cell Metabolism* 22, no. 1 (2015): 86–99.

Bridgeman, Brent. "A simple answer to a simple question on changing answers." *Journal of Educational Measurement* 49, no. 4 (2012): 467–68.

Carter, Evan C., et al. "A series of meta-analytic tests of the depletion effect: Self-control does not seem to rely on a limited resource." *Journal of Experimental Psychology: General* 144, no. 4 (2015): 796–815.

Cepeda, Nicholas J., et al. "Spacing effects in learning: A temporal ridgeline of optimal retention." *Psychological Science* 19, no. 11 (2008): 1095–102.

Chang, Y. K., et al. "The effects of acute exercise on cognitive performance: A meta-analysis." *Brain Research* 1453 (2012): 87–101.

Chiesa, A., et al. "Does mindfulness training improve cognitive abilities? A systematic review of neuropsychological findings." *Clinical Psychology Review* 31, no. 3 (2011): 449–64.

Christopher, Eddie A., and Jill Talley Shelton. "Individual differences in working memory predict the effect of music on student performance." *Journal of Applied Research in Memory and Cognition* 6, no. 2 (2017): 167–73.

Cousins, James N., et al. "Does splitting sleep improve long-term memory in chronically sleep deprived adolescents?" *npj Science of Learning* 4, no. 1 (2019): Art. No. 8.

Cowan, N. "The many faces of working memory and short-term storage." *Psychonomic Bulletin and Review* 24, no. 4 (2017): 1158–70.

Cowan, Nelson. "The magical number 4 in short-term memory: A reconsideration of mental storage capacity." *Behavioral and Brain Sciences* 24, no. 1 (2001): 87–114.

Cox, K. H., et al. "Investigation of the effects of solid lipid curcumin on cognition and mood in a healthy older population." *Journal of Psychopharmacology* 29, no. 5 (2015): 642–51.

Cutino, Chelsea M., and Michael A. Nees. "Restricting mobile phone access during homework increases attainment of study goals." *Mobile Media & Communication* 5, no. 1 (2016): 63–79.

D'Angiulli, Amedeo, et al. "Vividness of visual imagery and incidental recall of verbal cues, when phenomenological availability reflects long-term memory accessibility." *Frontiers in Psychology* 4 (2013): 1–18.

Dehaene, S., and J. P. Changeux. "Experimental and theoretical approaches to conscious processing." *Neuron* 70, no. 2 (2011): 200–27.

Dignath, Charlotte, and Gerhard Büttner. "Components of fostering self-regulated learning among students: A meta-analysis on intervention studies at primary and secondary school level." *Metacognition and Learning* 3, no. 3 (2008): 231–64.

Dik, Giel, and Henk Aarts. "Behavioral cues to others' motivation and goal pursuits: The perception of effort facilitates goal inference and contagion." *Journal of Experimental Social Psychology* 43, no. 5 (2007): 727–37.

Doran, George T. "There's a SMART way to write management's goals and objectives." *Management Review* 70, no. 11 (1981): 35–36.

Duckworth, Angela L., et al. "Self-control and academic achievement." *Annual Review of Psychology* 70, no. 1 (2019): 373–99.

Duckworth, Angela Lee, et al. "Self-regulation strategies improve self-discipline in adolescents: Benefits of mental contrasting and implementation intentions." *Educational Psychology* 31, no. 1 (2011): 17–26.

Dunlosky, John, et al. "Improving students' learning with effective learning techniques: Promising directions from cognitive and educational psychology." *Psychological Science in the Public Interest* 14, no. 1 (2013): 4–58.

Ericsson, K. Anders, and Robert Pool. *Peak: Secrets from the New Science of Expertise.* Boston, MA: Eamon Dolan/Houghton Mifflin Harcourt, 2016.

Fiebig, Florian, and Anders Lansner. "Memory consolidation from seconds to weeks: A three-stage neural network model with autonomous reinstatement dynamics." *Frontiers in Computational Neuroscience* 8 (2014): Art. No. 64, 1–17.

Fox, M. D., et al. "The human brain is intrinsically organized into dynamic, anticorrelated functional networks." *PNAS* 102 (2005): 9673–78.

Garcia-Argibay, Miguel, et al. "Efficacy of binaural auditory beats in cognition, anxiety, and pain perception: A meta-analysis." *Psychological Research* 83, no. 2 (2019): 357–72.

Garrison, Kathleen A., et al. "Meditation leads to reduced default mode network activity beyond an active task." *Cognitive, Affective, & Behavioral Neuroscience* 15, no. 3 (2015): 712–20.

Geng, J., et al. *Cochrane Database of Systematic Reviews,* no. 12 (2010): Art. No. CD007769.

Gervain, Judit, et al. "Valproate reopens critical-period learning of absolute pitch." *Frontiers in Systems Neuroscience* 7, no. 102 (2013): Art. No. 102.

Ghosh, VE, and Gilboa, A. "What is a memory schema? A historical perspective on current neuroscience literature." Neuropsychologia 53, (2014): 104–114.

Glade, M. J. "Caffeine—Not just a stimulant." *Nutrition* 26, no. 10 (2010): 932–38.

Gothe, Neha P., et al. "Yoga effects on brain health: A systematic review of the current literature." *Brain Plasticity* 5, no. 1 (2019): 105–22.

Handel, David. "How to unlock the amazing power of your brain and become a top student." *Medium* (2019). https://medium.com/better-humans/how-to-unlock-the-amazing-power-of-your-brain-and-become-a-top-student-369e5ba59484.

Harvard Medical School. "Blue light has a dark side." *Harvard Health Letter* (2012, updated 2018). https://www.health.harvard.edu/staying-healthy/blue-light-has-a-dark-side.

Haskell, C. F., et al. "Behavioural effects of compounds co-consumed in dietary forms of caffeinated plants." *Nutrition Research Reviews* 26, no. 1 (2013): 49–70.

Heisz, J. J., et al. "The effects of physical exercise and cognitive training on memory and neurotrophic factors." *Journal of Cognitive Neuroscience* 29, no. 11 (2017): 1895–907.

Himmer, L., et al. "Rehearsal initiates systems memory consolidation, sleep makes it last." *Science Advances* 5, no. 4 (2019): eaav1695.

Hofmann, Wilhelm, et al. "Yes, but are they happy? Effects of trait self-control on affective well-being and life satisfaction." *Journal of Personality* 82, no. 4 (2014): 265–77.

Hruby, George G., and Usha Goswami. "Neuroscience and reading: A review for reading education researchers." *Reading Research Quarterly* 46, (2011): 156–72.

Hughes, Nicola, and Jolanta Burke. "Sleeping with the frenemy: How restricting 'bedroom use' of smartphones impacts happiness and wellbeing." *Computers in Human Behavior* 85, (2018): 236–44.

Hulleman, Chris S., et al. "Enhancing interest and performance with a utility value intervention." *Journal of Educational Psychology* 102, no. 4 (2010): 880–95.

Jansen, Renée S., et al. "An integrative review of the cognitive costs and benefits of note-taking." *Educational Research Review* 22 (2017): 223–33.

Jenkins, E. M., et al. "Do stair climbing exercise 'snacks' improve cardiorespiratory fitness?" *Applied Physiology, Nutrition, and Metabolism* 44, no. 6 (2019): 681–84.

Josselyn, Sheena A., and Paul W. Frankland. "Memory allocation: Mechanisms and function." *Annual Review of Neuroscience* 41, no. 1 (2018): 389–413.

Jwa, Anita. "DIY tDCS: A need for an empirical look." *Journal of Responsible Innovation* 5, no. 1 (2018): 103–8.

Kang, S., and T. R. Kurtzberg. "Reach for your cell phone at your own risk: The cognitive costs of media choice for breaks." *Journal of Behavioral Addictions* 8, no. 3 (2019): 395–403.

Kapadia, Chaitali, and Shimul Melwani. "More tasks, more ideas: The positive spillover effects of multitasking on subsequent creativity." *Journal of Applied Psychology* (2020): Advance publication online.

Karpicke, J. D., and J. R. Blunt. "Retrieval practice produces more learning than elaborative studying with concept mapping." *Science* 331, no. 6018 (2011): 772–75.

Karpicke, Jeffrey D. "Retrieval-based learning: Active retrieval promotes meaningful learning." *Current Directions in Psychological Science* 21, no. 3 (2012): 157–63.

Kiewra, Kenneth A., et al. "Note-taking functions and techniques." *Journal of Educational Psychology* 83, no. 2 (1991): 240–45.

Kornell, Nate, and Robert A. Bjork. "Learning concepts and categories: Is spacing the 'enemy of induction'?" *Psychological Science* 19, no. 6 (2008): 585–92.

Kornell, Nate, Matthew J. Hays, and Robert A. Bjork. "Unsuccessful retrieval attempts enhance subsequent learning." *Journal of Experimental Psychology: Learning, Memory, and Cognition* 35, no. 4 (2009): 989–98.

Kühn, Simone, et al. "The importance of the default mode network in creativity—a structural MRI study." *Journal of Creative Behavior* 48, no. 2 (2014): 152–63.

Kuznekoff, Jeffrey H., and Scott Titsworth. "The impact of mobile phone usage on student learning." *Communication Education* 62, no. 3 (2013): 233–52.

Lally, Phillippa, et al. "How are habits formed: Modelling habit formation in the real world." *European Journal of Social Psychology* 40, no. 6 (2010): 998–1009.

Laws, Keith R., et al. "Is ginkgo biloba a cognitive enhancer in healthy individuals? A meta-analysis." *Human Psychopharmacology: Clinical and Experimental* 27, no. 6 (2012): 527–33.

Leinenger, Mallorie. "Phonological coding during reading." *Psychological Bulletin* 140, no. 6 (2014): 1534–55.

Leroy, Sophie, and Theresa M. Glomb. "Tasks interrupted: How anticipating time pressure on resumption of an interrupted task causes attention residue and low performance on interrupting tasks and how a 'ready-to-resume' plan mitigates the effects." *Organization Science* 29, no. 3 (2018): 380–97.

Liles, Jenny, et al. "Study habits of medical students: An analysis of which study habits most contribute to success in the preclinical years." *MedEdPublish* 7, no. 1 (2018): 61.

Lu, Bai, et al. "BDNF-based synaptic repair as a disease-modifying strategy for neurodegenerative diseases." *Nature Reviews: Neuroscience* 14, no. 6 (2013): 401–16.

Lu, Jackson G., et al. "'Switching on' creativity: Task switching can increase creativity by reducing cognitive fixation." *Organizational Behavior and Human Decision Processes* 139 (2017): 63–75.

Ly, C., et al. "Psychedelics promote structural and functional neural plasticity." *Cell Reports* 23, no. 11 (2018): 3170–82.

Lyons, I. M., and S. L. Beilock. "When math hurts: Math anxiety predicts pain network activation in anticipation of doing math." *PLOS One* 7, no. 10 (2012): e48076.

Macedonia, M., et al. "Depth of encoding through observed gestures in foreign language word learning." *Frontiers in Psychology* 10 (2019): Art. No. 33.

Madjar, Nora, and Christina E. Shalley. "Multiple tasks' and multiple goals' effect on creativity: Forced incubation or just a distraction?" *Journal of Management* 34, no. 4 (2008): 786–805.

Mark, Gloria, et al. "How blocking distractions affects workplace focus and productivity." In *Proceedings of the 2017 ACM International Joint Conference on Pervasive and Ubiquitous Computing and Proceedings of the 2017 ACM International Symposium on Wearable Computers,* 928–34: ACM, 2017.

Mark, Gloria, et al. "Neurotics can't focus: An in situ study of online multitasking in the workplace." In *Proceedings of the 2016 CHI Conference on Human Factors in Computing Systems,* 1739–44: ACM, 2016.

Mattson, M. P. "An evolutionary perspective on why food overconsumption impairs cognition." *Trends in Cognitive Science* 23, no. 3 (2019): 200–12.

Mayer, Richard E. *The Cambridge Handbook of Multimedia Learning.* 2nd ed. New York, NY: Cambridge University Press, 2014.

Medeiros-Ward, N., et al. "On supertaskers and the neural basis of efficient multitasking." *Psychonomic Bulletin & Review* 22, no. 3 (2015): 876–83.

Miller, Marshall, et al. "Role of fruits, nuts, and vegetables in maintaining cognitive health." *Experimental Gerontology* 94 (2017): 24–28.

Milyavskaya, Marina, and Michael Inzlicht. "What's so great about self-control? Examining the importance of effortful self-control and temptation in predicting real-life depletion and goal attainment." *Social Psychological and Personality Science* 8, no. 6 (2017): 603–11.

Moffitt, Terrie E., et al. "A gradient of childhood self-control predicts health, wealth, and public safety." *PNAS* 108, no. 7 (2011): 2693–98.

Mokhtari, Kouider, et al. "Connected yet distracted: Multitasking among college students." *Journal of College Reading and Learning* 45, no. 2 (2015): 164–80.

Molenberghs, Pascal, et al. "Neural correlates of metacognitive ability and of feeling confident: A large-scale fMRI study." *Social Cognitive and Affective Neuroscience* 11, no. 12 (2016): 1942–51.

Nehlig, A. "Is caffeine a cognitive enhancer?" *Journal of Alzheimer's Disease* 20, suppl. 1 (2010): S85–S94.

Nestor, J. *Breath: The New Science of a Lost Art.* New York, NY: Riverhead Books, 2020.

O'Connor, Anahad. "How the hum of a coffee shop can boost creativity." *New York Times,* June 21, 2013.

Oakley, Barbara A., and Terrence J. Sejnowski. "What we learned from creating one of the world's most popular MOOCs." *npj Science of Learning* 4 (2019): Art. No. 7.

Oakley, Barbara, et al. *Uncommon Sense Teaching.* New York, NY: Penguin Random House, 2021.

Oettingen, Gabriele, and Klaus Michael Reininger. "The power of prospection: Mental contrasting and behavior change." *Social and Personality Psychology Compass* 10, no. 11 (2016): 591–604.

Pan, Steven C., and Robert A. Bjork. "Chapter 11.3 Acquiring an accurate mental model of human learning: Towards an owner's manual." In *Oxford Handbook of Memory, Vol. II: Applications.* In press.

Poo, M. M., et al. "What is memory? The present state of the engram." *BMC Biology* 14 (2016): Art. No. 40.

Pribis, Peter, and Barbara Shukitt-Hale. "Cognition: The new frontier for nuts and berries." *American Journal of Clinical Nutrition* 100, 1 (2014): 347S–352S.

Rayner, Keith, et al. "So much to read, so little time: How do we read, and can speed reading help?" *Psychological Science in the Public Interest* 17, no. 1 (2016): 4–34.

Rendeiro, C., et al. "Flavonoids as modulators of memory and learning: Molecular interactions resulting in behavioural effects." *Proceedings of the Nutritional Society* 71, no. 2 (2012): 246–62.

Renno-Costa, C., et al. "Computational models of memory consolidation and long-term synaptic plasticity during sleep." *Neurobiology of Learning and Memory* 160 (2019): 32–47.

Repantis, Dimitris, et al. "Modafinil and methylphenidate for neuroenhancement in healthy individuals: A systematic review." *Pharmacological Research* 62, no. 3 (2010): 187–206.

Rittle-Johnson, Bethany, et al. "Not a one-way street: Bidirectional relations between procedural and conceptual knowledge of mathematics." *Educational Psychology Review* 27, no. 4 (2015): 587–97.

Roediger III, Henry L., and Jeffrey D. Karpicke. "Test-enhanced learning: Taking memory tests improves long-term retention." *Psychological Science* 17, no. 3 (2006): 249–55.

Roosevelt, Theodore. "The books that I read and when and how I do my reading." *Ladies' Home Journal* 32, no. 4 (1915). https://www.theodorerooseveltcenter .org/Research/Digital-Library/Record/ImageViewer?libID=o292909&image No=1.

Rowland, C. A. "The effect of testing versus restudy on retention: A meta-analytic review of the testing effect." *Psychology Bulletin* 140, no. 6 (2014): 1432–63.

Rubinstein, Joshua S., et al. "Executive control of cognitive processes in task switching." *Journal of Experimental Psychology: Human Perception and Performance* 27, no. 4 (2001): 763–97.

Ryan, Richard M., and Edward L. Deci. "Self-determination theory and the facilitation of intrinsic motivation, social development, and well-being." *American Psychologist* 55, no. 1 (2000): 68.

Schapiro, Anna C., et al. "Complementary learning systems within the hippocampus: A neural network modelling approach to reconciling episodic memory with statistical learning." *Philosophical Transactions of the Royal Society B: Biological Sciences* 372 (2017). https://doi.org/10.1098/rstb.2016.0049.

Sedlmeier, Peter, et al. "The psychological effects of meditation: A meta-analysis." *Psychological Bulletin* 138, no. 6 (2012): 1139–1171.

Sekeres, M. J., et al. "The hippocampus and related neocortical structures in memory transformation." *Neuroscience Letters* 680 (2018): 39–53.

Sekeres, Melanie J., et al. "Mechanisms of memory consolidation and transformation." In *Cognitive Neuroscience of Memory Consolidation,* 17–44. Switzerland: Springer International Publishing, 2017.

Sheeran, Paschal, et al. "The interplay between goal intentions and implementation intentions." *Personality and Social Psychology Bulletin* 31, no. 1 (2005): 87–98.

Shirota, Y., et al. "Neuroscientists do not use non-invasive brain stimulation on themselves for neural enhancement." *Brain Stimulation* 7, no. 4 (2014): 618–19.

Sinanaj, I., et al. "Neural underpinnings of background acoustic noise in normal aging and mild cognitive impairment." *Neuroscience* 310 (2015): 410–21.

Sio, U. N., and T. C. Ormerod. "Does incubation enhance problem-solving? A meta-analytic review." *Psychological Bulletin of Science, Technology & Society* 135, no. 1 (2009): 94–120.

Smith, Amy M., et al. "Retrieval practice protects memory against acute stress." *Science* 354, no. 6315 (2016): 1046–48.

Smith, M. A., et al. "Glucose enhancement of human memory: A comprehensive research review of the glucose memory facilitation effect." *Neuroscience & Biobehavioral Reviews* 35, no. 3 (2011): 770–83.

Smith, M. E., and M. J. Farah. "Are prescription stimulants 'smart pills'? The epidemiology and cognitive neuroscience of prescription stimulant use by normal healthy individuals." *Psychological Bulletin* 137, no. 5 (2011): 717–41.

Sniehotta, Falko F., et al. "Action plans and coping plans for physical exercise: A longitudinal intervention study in cardiac rehabilitation." *British Journal of Health Psychology* 11, no. 1 (2006): 23–37.

Socci, V., et al. "Enhancing human cognition with cocoa flavonoids." *Frontiers in Nutrition* 4 (2017): Art. No. 10.

Standing, Lionel, et al. "Perception and memory for pictures: Single-trial learning of 2500 visual stimuli." *Psychonomic Science* 19, no. 2 (1970): 73–74.

Stork, Matthew J., et al. "Let's go: Psychological, psychophysical, and physiological effects of music during sprint interval exercise." *Psychology of Sport and Exercise* 45 (2019): 101547.

Straube, B., et al. "Memory effects of speech and gesture binding: Cortical and hippocampal activation in relation to subsequent memory performance." *Journal of Cognitive Neuroscience* 21, no. 4 (2009): 821–36.

Sweller, John, et al. *Cognitive Load Theory.* New York, NY: Springer-Verlag, 2011.

Szuhany, Kristin L., et al. "A meta-analytic review of the effects of exercise on brain-derived neurotrophic factor." *Journal of Psychiatric Research* 60 (2015): 56–64.

Thompson, Derek. "A formula for perfect productivity: Work for 52 Minutes, Break for 17." *The Atlantic,* September 17, 2014. https://www.theatlantic.com/business/archive/2014/09/science-tells-you-how-many-minutes-should-you-take-a-break-for-work-17/380369/.

Tobin, K. J. "Fast-food consumption and educational test scores in the USA." *Child: Care, Health and Development* 39, no. 1 (2013): 118–24.

Treadway, Michael T., et al. "Dopaminergic mechanisms of individual differences in human effort-based decision-making." *Journal of Neuroscience* 32, no. 18 (2012): 6170–76.

Turow, Gabe, and James D. Lane. "Binaural beat stimulation: Altering vigilance and mood states." In *Music, Science, and the Rhythmic Brain: Cultural and Clinical Implications,* 122–39. New York, NY: Routledge, 2011.

U.S. Department of Health and Human Services. "Physical Activity Guidelines for Americans, 2nd edition" (2018). https://health.gov/paguidelines/second-edition/pdf/Physical_Activity_Guidelines_2nd_edition.pdf.

Ullman, Michael T., and Jarrett T. Lovelett. "Implications of the declarative/procedural model for improving second language learning: The role of memory enhancement techniques." *Second Language Research* 34, no. 1 (2016): 39–65.

van Kesteren, Marlieke Tina Renée, and Martijn Meeter. "How to optimize knowledge construction in the brain." *npj Science of Learning* 5, Article number: 5 (2020).

van Praag, Henriette. "Exercise and the brain: Something to chew on." *Trends in Neurosciences* 32, no. 5 (2009): 283–90.

Walker, Matthew. *Why We Sleep: The New Science of Sleep and Dreams.* New York, NY: Penguin, 2017.

Wamsley, Erin J. "Memory consolidation during waking rest." *Trends in Cognitive Sciences* 23, no. 3 (2019): 171–73.

Wamsley, Erin J., et al. "Dreaming of a learning task is associated with enhanced sleep-dependent memory consolidation." *Current Biology* 20, no. 9 (2010): 850–55.

Ward, Adrian F., et al. "Brain drain: The mere presence of one's own smartphone reduces available cognitive capacity." *Journal of the Association for Consumer Research* 2, no. 2 (2017): 140–54.

Wardle, Margaret C., et al. "Amping up effort: Effects of d-amphetamine on human effort-based decision-making." *Journal of Neuroscience* 31, no. 46 (2011): 16597–602.

Winne, Philip H., and Allyson F. Hadwin. "Studying as self-regulated learning." In *Metacognition in Educational Theory and Practice,* edited by D. Hacker et al., 27–30. Mahwah, NJ: Lawrence Erlbaum Associates, 1998.

Winter, Lloyd Bud. *Relax and Win: Championship Performance in Whatever You Do.* San Diego, CA: Oak Tree Publications, 1981.

Xie, Lulu, et al. "Sleep drives metabolite clearance from the adult brain." *Science* 342, no. 6156 (2013): 373–77.

Yang, Guang, et al. "Sleep promotes branch-specific formation of dendritic spines after learning." *Science* 344, no. 6188 (2014): 1173–78.

Zureick, A. H., et al. "The interrupted learner: How distractions during live and video lectures influence learning outcomes." *Anatomical Sciences Education* 11, no. 4 (2018): 366–76.

Index

Page numbers in *italics* refer to illustration captions.

About the Authors

Rachel Oakley

BARBARA OAKLEY, PhD, PE is a professor of engineering at Oakland University and was named Michigan's Distinguished Professor of the Year in 2018. Oakley is also a popular instructor of massive open online courses and has taught millions how to learn more effectively.

Ofu Yu

OLAV SCHEWE is the founder and CEO of Educas, an educational technology start-up that develops solutions to help students learn. He is also an educational consultant for one of the world's largest educational tech companies, Kahoot!.